HOW NON-TECHNICAL FOUN
PROFITABLE, SCALABLE S

"...true builder's guide to profitability and market fit, full of practical advice and real examples of success stories..."

– **Jolie Miller,** *Director of Content at LinkedIn*

THE 6 STARTUP STAGES

Jason Hishmeh   Stas Chernychko   6startupstages.com

THE **6 STARTUP**™
STAGES

The 6 Startup Stages

Jason Hishmeh   *Stas Chernychko*

6startupstages.com

Table of Contents

Introduction

Reviews

⭐ ⭐ ⭐ ⭐ ⭐

"In The Six Startup Stages, you'll learn why so many ventures fail. The insight: founders often overestimate their progress and overinvest, burning through capital before they realize they are off course. With rich examples, the book shows how to avoid that fate by making the right moves for your venture's true stage of development."

Thomas Eisenmann

Author of "Why Startups Fail" and Professor, Harvard Business School

⭐ ⭐ ⭐ ⭐ ⭐

"The 6 Startup Stages is a true builder's guide to profitability and market fit, full of practical advice and real examples of success stories. Perfect blend of reality and optimism to help you build companies that can weather and thrive amidst changing market dynamics."

Jolie Miller

Director of Content, LinkedIn

⭐ ⭐ ⭐ ⭐ ⭐

"This book might just save your business. Having gone through a few failed start-ups, I wish I would have had this book to stop me from making some bad decisions. Jason and Stas have put together a powerful step-by-step guide to start-up success. A must read for founders."

Brant Menswar

Best-Selling Author and Founder of Black Sheep Foundry

★ ★ ★ ★ ★

"Jason and Stas have written the book needed by entrepreneurs looking to answer the most important questions faced by startups..."

Dave Jaworski
Principal Product Manager, Microsoft 365 Copilot and Microsoft Teams, Author of "Microsoft Secrets", and Co-founder of Tools of the Trade Media, LLC and DirectorPrep.com

★ ★ ★ ★ ★

"The Six Startup Stages guide entrepreneurs through a practical step-wise and iterative path to building a business, helping them avoid many pitfalls and blind spots along the way..."

Menachem Tabanpour
Managing Director of gener8tor Luxembourg Investment Accelerator

★ ★ ★ ★ ★

"6 Startup Stages' is an essential roadmap for any entrepreneur. It's a micro-MBA with actionable insights and tools that are crucial for navigating the uncertain journey of building a startup. Avoid common pitfalls with practical real-world examples so you can turn your vision into a successful business."

John Eng
Funding Ecosystem Partner of Right Side Capital Management

★ ★ ★ ★ ★

"It's equally beneficial for non-founders, providing valuable insights into the journey of launching a company and the essential steps required for success!"

Peggy Tsai
Co-Author of "The AI Book", Chief Data Officer at BigID, Tech Start-up Advisor

★ ★ ★ ★ ★

"This insightful roadmap is brimming with practical techniques and tools to vet your ideas and lay the foundation for a bright, bold future."

Karin Hurt
CEO of Let's Grow Leaders and author of Courageous Cultures: How to Build Teams of Micro-Innovators

About the Authors

I started my first entrepreneurial adventure while attending University in my 20s— a business to help inventors leverage dormant patents they owned and turn them into profitable companies. That business completely failed.

I did not know anything about starting a business, getting customers, marketing, building a business, or how business worked in general. I did not even know how to work with other people effectively. I also did not know how costly it would be to launch a business. I could not even afford the costs to get it fully started and operational.

That experience, along with many other successes and failures over the past decades, helped me learn lessons and acquire the skills I use to this day to build and run successful companies.

Many years later, after other entrepreneurial endeavors, selling one company, and experiences working at Fortune 500 companies, my current partner Stas and I met and formed our current business called Varyence.

Varyence helps non-technical founders of early-stage SaaS (Software-as-a-Service) companies launch and scale, with a focus on profitability. We have helped multiple startups grow to multi-millions in revenue, with one company even being acquired by a public company.

In this book, we share stories, insider playbooks, and tips that come from real-world experiences Stas and I have had, building successful companies.

We are not coaches or influencers. Our focus is not on selling books. We do not believe in encouraging founders to raise millions from venture capitalists in hopes of turning their startup into a unicorn. We do not believe in giving advice to a new startup founder about how to get to $10m in annual revenue when they have not even generated $1 in revenue.

We also do not believe in living flashy lives or trying to impress people - in fact, most of our employees drive much nicer cars than we do. Most entrepreneurs spend many years working hard and making personal sacrifices to build their companies.

We encourage founders not to get caught up with announcements they see on social media about startups raising millions in funding or cashing out after a few months. Raising money is not the goal; building successful, profitable companies is the goal. In fact, raising money, or getting involved with the wrong investors, could actually hurt your chances of success.

Our focus is on launching and building realistically attainable, profitable, successful companies. Companies that help transform the world through innovative technology solutions. This is what we love doing every day. If we can help a founder build a profitable business that achieves even $5-$10 million in annual revenue, for us this is a success.

Why We Wrote This Book

We know what it is like to feel like nothing is going right. To feel like everything is failing. To wonder how you are going to make payroll. To feel like there is no one to count on to help you out. And to sometimes feel hopeless, alone, and ready to give up.

Recently I spoke with the CEO of a startup who told me they are shutting down their company. They started their company 6 months prior and had spoken to some very prominent Venture Capitalists (VC) who told them their business was very interesting, was VCworthy for funding, and that they planned to invest in it. This founder did everything they could to position themselves for that investment. They eventually ran out of capital, went back to those VCs and were then told they were too early for investments. This situation happens more than you would imagine. Founders may not have had enough experience to realize there are miles of distance between an investor verbally saying they are interested in investing and then actually investing. Founders may also not realize there are miles of distance between them achieving the unspoken traction the investor needs to see to truly invest in their company.

Although many founders fail and get into these traps, it does not need to turn out that way. They do not need to burn through all their capital and toss out their dreams. They do not need to jeopardize their family, their future, and their finances.

We know with the right knowledge, processes, and guidance, founders can navigate each stage of the startup lifecycle and avoid common missteps along the way.

As a company, we have over a decade of experience launching tech startups. We see founders make a lot of the same mistakes over and over. We have personally been through each stage of the startup lifecycle multiple times, made lots of mistakes ourselves, and learned from those mistakes. We have many battle wounds to show for it. We still make mistakes and still learn painful lessons, but hopefully what we have already learned can help you avoid making the same mistakes we made.

I recently shared with a founder that being an entrepreneur can feel like an emotional roller coaster. Some days you can feel on top of the world, and some days, you can feel like a complete loser and want to give up. Try living that emotional roller coaster for years or decades and see how it impacts your life and your family. We know what it is like because we live it.

We know our guidance can help founders avoid:

- Building products nobody wants.
- Marketing products to the wrong audience.
- Running out of capital.

- Losing out on their dream.
- And running out of hope…

Over the years we created internal playbooks to help us guide non-technical CEOs of our portfolio companies and of our commercial clients.

In this book, we share our insider tips that we have seen first-hand help founders succeed.

We focus on key learnings non-technical founders need to know so they can:

- Approach each startup stage like a more seasoned startup founder.
- Avoid costly mistakes.
- Build businesses that are attractive not only to investors but to paying customers.
- Create opportunities to exit or get acquired.
- Obtain financial freedom for them and their family.
- Keep persisting and adapting until they are successful…

We wrote this book to help other founders and to share information that we know can help them succeed and build their dreams.

What Most Startups Get Wrong

Less than 4% of all startups attain even $1 million in annual revenue[1], and 90% of businesses worldwide fail. Yet, in the US alone every year, over 4,000,000 new businesses are launched.[2]

With all the information available to people these days, you may wonder, "Why do founders not just find out the best way to create successful companies and do it that way?" "Why do they keep failing?"

---

[1] BusinessKnowHow.com survey
[2] US Census Bureau Business Formation Statistics. Published by Economic Innovation Group. January 19, 2022

We wondered the same thing. We studied our portfolio companies and commercial clients we advised, and we started to understand why certain startups succeeded or failed. External investors also approached us—curious why our portfolio companies and companies we helped guide, have a much greater success rate than the industry average. We were surprised by some of our findings about why founders kept failing.

Some of our findings were that many founders:

- Believe they already have the most important angles covered and think they know how to avoid common pitfalls.

- Do not realize what is holding them back.

- Are missing key pieces of information or experience to get them to the next level.

- Do not realize the nuances of launching and scaling a successful tech startup that experienced founders know.

- Think they are in a different stage than they really are, causing them to focus on the wrong thing at the wrong time.

Founders can often be their own worst enemy, blocking their own success and not realizing where they are failing, before it is too late. Sometimes it is due to personal blindspots, sometimes it is due to stubbornness, and often it is because they just do not have the right guidance, leadership team, or advisors around to support them.

> **TIP**
>
> "Many founders fail because they do not believe they have any blind spots. And once they do realize it, it is often too late."

In this book, we share our insights, insider playbooks, and lessons we have learned. We call this The 6 Startup Stages.

Why Does This Matter?

Until you are successful, the expectation from the world is that your startup is going to fail, you are going to go out of business, and you are going to burn through capital with nothing to show for it. And there is a lot of data to back up those expectations.

At one point, my father told me to go get a good job so that I would not end up destitute. He had a job, a pension, and a retirement plan. He could not understand why I wanted to take such risks to build a business. When your father tells you something like that, it is not easy to forget it. The decision to be an entrepreneur is not something to take lightly. You really are risking a stable career, your family's finances, your reputation, and your future. And there is a high probability you will fail. So in many ways, my father was not wrong; it is a real risk.

It is important to prepare yourself for this journey. It is important to get into the right mindset and get the right guidance to increase your chances of success, avoid missteps, and create a successful business. Even with the best advice, you will still have setbacks. But you will be better prepared for those setbacks. And you will learn from them. And you will realize those setbacks are required steps before you can be successful.

Understanding the concepts in this book can mean the difference between success and failure and put you steps ahead of other startups competing for the same customers, funding, and talent.

Founders who leverage The 6 Startup Stages framework are in a better position to create in-demand products, get to market more quickly, raise the right amount of capital, reach profitability, and build successful companies.

These concepts can mean the difference between bankruptcy and building a legacy for your family. These concepts can mean the difference between losing hope, and persisting long enough to see your dream become a reality.

Who Should Read This Book?

This book is primarily written for non-technical founders who are launching or scaling SaaS startups.

The fundamentals of The 6 Startup Stages can help you at every stage of your startup journey. While no two startups are the same, our framework helps non-technical startup founders understand the fundamental principles of transforming their startup idea into a profitable, sustainable business.

For more tools, resources, guidance, and details, we recommend you check out the "Bonus Tools and Content" section. If you are not sure about a startup term used in this book, check out the Glossary at the end for the top 100 terms we suggest you familiarize yourself with. You can also find a link in the bonus section to take a free online quiz to identify where you are at and get personalized recommendations for your situation.

About The 6 Startup Stages

Regardless of where you are in the process, we know you can benefit from the guidance we share in The 6 Startup Stages.

Let us start with a brief overview of the 6 stages of a startup. We will then go into each of these stages in detail in subsequent chapters.

1. Idea Stage:

This first stage focuses on developing and validating your startup idea. More specifically, we identify a problem that needs solving and a viable product solution. We recommend founders set aside one month at this stage, refining the problem they are trying to solve and defining how their idea solves it. This stage should not require a lot of expenses. Additionally, most of the effort here is completed by the founders.

⚖️ 2. Validation Stage:

This stage focuses on your business model and product validation. During validation, entrepreneurs refine their ideas with real-world feedback and identify if or why consumers would be willing to pay for their solution. We recommend founders set aside one to several months for this stage. More complex startups could require up to a year for this stage. This stage is generally bootstrapped by the founders and the costs at this stage should also not be substantial.

🚀 3. Launch Stage:

This stage focuses on putting an MVP (Minimal Viable Product) into early adopters' hands, balancing time and resources. Depending on the type of product, we recommend founders set aside between one to three months for this stage. If your launch is projected to take more than six months, you may need to reduce the scope of your initial MVP.

📣 4. Market Fit Stage:

This stage focuses on pursuing product-market fit and delivering your first official product adapted from your MVP. This stage of development and growth could take anywhere from 3 to 12 months, depending on the scope of your product. It is important to continue refining your product (based on customer feedback) enough to achieve product-market fit without spending so much time that you cause major delays in getting your product to market. For example, spending multiple years iterating on your first official product to get it "just right" could cause you to lose money, lose momentum, get overtaken by competitors, and cause early adopters to give up on your product.

◎ 5. Market Focus Stage:

At this stage, a startup has achieved a significant level of growth and is expanding its market share. The Market Focus stage objective is increased adoption of your product within your target audience without digressing from your audience or expanding your market beyond your Ideal Customer Profile (ICP). It also emphasizes improving the health and resiliency of the company. This stage can often take one to three years.

↗ 6. Scale Stage:

This stage focuses on major expansions of revenue, reducing organizational debt, technical debt, and fiscal debt. It is also important at this stage to improve the long-term sustainability and resiliency of your company. Often, at this stage, a company may expand the scope of its ICP (Ideal Customer Profile) or add additional ones. This stage is ongoing until you exit the company.

Below is a visualization of these stages and timelines.

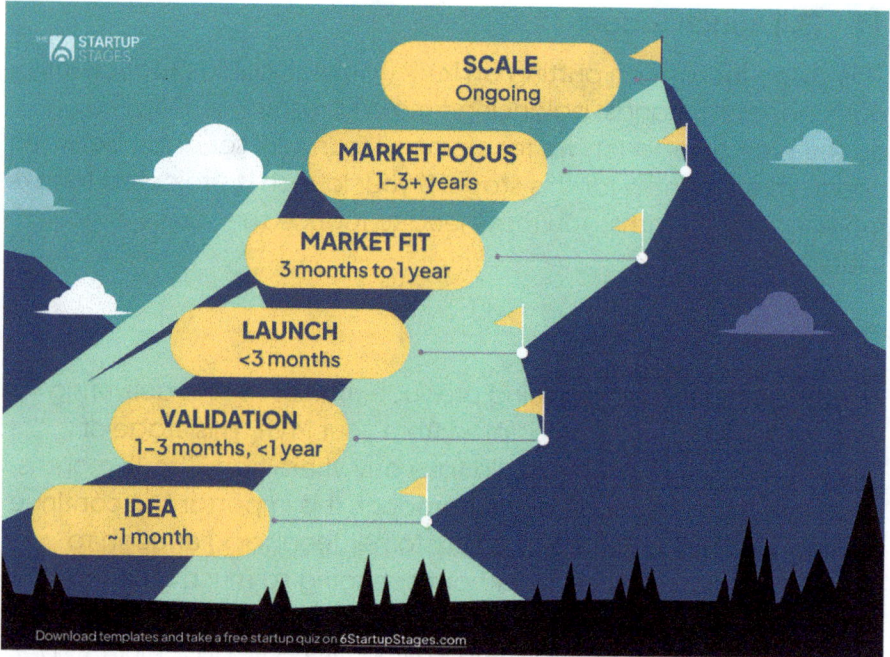

STARTUP STAGES

SCALE
Ongoing

MARKET FOCUS
1–3+ years

MARKET FIT
3 months to 1 year

LAUNCH
<3 months

VALIDATION
1–3 months, <1 year

IDEA
~1 month

Download templates and take a free startup quiz on 6StartupStages.com

Breaking down the startup journey into these stages can help you focus your energy and avoid getting overwhelmed. More importantly, it lays the foundation for avoiding the issues and challenges that cause startups to fail. We have seen many entrepreneurs struggle to move from one stage to the next or completely skip stages due to a lack of awareness. We have been guilty of doing the same in the past.

We know that following our framework can help you save time and money, hone your focus, and ultimately build a successful, sustainable business.

01

The Idea Stage:
FROM PROBLEM TO VISION

1 The Idea Stage: From Problem to Vision

Early in my career, I tried many tech startup ideas. Some I wish I never started, and some I wish I never stopped.

For example, I was trying to build an online time tracking software at one point and an online accounting platform at another point. I remember sitting at a restaurant with a friend I admired for his entrepreneurial experience.

He told me the online accounting system is a stupid idea and will never work. I just accepted his advice and dropped the idea. I just assumed he knew from his experience it was a bad idea.

There were many times I also thought something was a bad idea. When I lived in NYC and Uber became popular, I couldn't get why it was a good idea. After all, I would just go down to the street, raise my hand, and get a yellow taxi. It would have taken me longer to use an app to request an Uber. Obviously, I got that wrong.

The main point is I had no framework for how to validate if an idea was good or not, whether it was my idea or someone else's. It took me years to figure out how to better evaluate business ideas.

For both of my ideas mentioned in the lesson above, if only I could have seen the future potential. If only I had known how to turn them into successful companies. If only...

But I did not. I was not aware of how to validate what should be built, how to acquire customers, how to build a leadership team, or what the first step really was in the process.

As a techie, it is fun to jump straight into building the product, thinking I know all the features a potential customer would love. For me, developing the tech is the fun part. Taking time to methodologically work step-by-step (refining the idea, performing market research, talking to potential customers, etc.) was never as exciting to me as jumping right into building the product because my strength was in the building part. I also had no idea how to do the other parts, so that was boring to me. I would skip from preliminary idea to launch, time and time again!

It took me years to learn this was not the right approach to achieve real results. Now, after decades of experience launching tech startups, I am more patient and work through each of The 6 Startup Stages listed in this book.

Whether you are an expert in your field or someone with a knack for problem-solving, do not skip the steps we outline in this Idea Stage. If you are not familiar with any terms you encounter in this book, do not worry; just flip to the Glossary at the end of the book.

This chapter will guide you through each step of the Idea Stage. Not only on things you should do but also things not to do this early in the process.

This chapter will explain how to:

- Define the problem you are trying to solve.
- Refine your product idea and solution for that problem.
- Hone your target audience and niche.
- Draft your business model.

As you spend time working through the Idea Stage, you will become more prepared to answer questions from potential investors and potential customers.

A well-defined idea will help you hone in on your niche market. However, keep in mind that the Idea Stage is about laying a good foundation, not getting everything right or getting everything done.

The Idea Stage will save you time and money, help you avoid some of the most common startup mistakes, build a convincing pitch for fundraising, and craft brand communications that resonate with potential customers.

🎓 LESSONS FROM REAL SITUATIONS

When I was at a recent Techstars Startup Weekend in the United States, there was one team leader who refused to take any input. They did not want to pivot or refine their idea. They did not want to adjust their approach.

One by one, each of their team members quit and joined another team, discouraged by the leader's inability to accept any input or adjust their vision. The team leader was so emotionally attached to their vision that they lost sight of how to accept feedback, refine their idea, and pivot their plans.

It is important to remain flexible and open to feedback. It is important to not get too emotional about feedback people give you.

Know that things will change and evolve as you go through this process. Remain flexible and adapt as you learn new information. Do not get too emotionally attached to an idea or approach. Keep open to input and advice from people.

Defining a Problem

Great startups begin with an idea of how to solve a problem. Many startup founders struggle to clearly communicate the problem their idea is trying to solve. While their messaging makes sense to them, very often it does not make sense to a potential customer.

For example, a founder might have a marketing website say something vague such as *"Helping make the world a better place for people who have diabetes."* A startup can not afford to have vague communication. Potential customers need to quickly understand what your business is offering and how it helps them.

I review hundreds of pitch decks per year. If I cannot understand what problem is being solved, I start to tune out because I realize the founder is not communicating well enough.

I also encounter many situations where founders create solutions and then try to find a problem they solve for. They start the process in reverse and waste time and money going in the wrong direction.

These disconnects can make it difficult or impossible to secure funding or get your target audience to understand why they should care about your product. This is a common problem at the Idea Stage and takes time to work through.

It is important to:

- Identify a real problem you are solving.
- Find the right target audience and partners.
- Focus on solving real needs of your target audience.
- Be able to communicate how your solution solves your target audience's problem.

When defining a problem, it is important to consider why people would pay for your solution. What value does your solution bring? How does it differ from current market offerings? Does it have a long-term use?

Let us walk through the process of defining a problem:

Start by writing a simple **problem statement**. We recommend using the following as a guide:

> *[background or context], the [target audience] is facing [specific problem]. This situation results in [negative effects]. There is a need to address it to achieve [desired outcome].*

Let us use an example of a healthcare tech startup.

> *Healthcare facilities in the United States struggle to access and utilize all necessary patient medical information because data is stored in disparate systems, disorganized, and not all digitized. This situation compromises treatments, frustrates medical providers, wastes patient and provider time, and increases costs for patients and the healthcare industry. There is a need to improve the efficiency and effectiveness of the healthcare system through better centralization, access, and organization of healthcare data.*

Once you understand each aspect of the problem, you can break it down into a clear set of goals and objectives, outlining the importance, urgency, relevance, benefit, constraints, assumptions, and dependencies of your proposed solution.

Continuing with our healthcare tech startup example:

Importance and Urgency:

	Description	Example
Relevance	Why is solving this problem relevant now?	The current state of healthcare records is riddled with problems and inefficiencies. New government regulations recently passed, mandate improved approaches to managing and sharing electronic health records.
Benefits	What are the potential benefits of solving the problem or seizing the opportunity?	The potential benefits include improved patient care, reduced medical errors, enhanced efficiency, and cost savings.
Urgency	How urgent is it to address this problem/opportunity?	Patient expectations are higher than ever, and healthcare costs continue to increase. People cannot afford healthcare. And recent regulations require improvements over the next several years.

Assumptions, Constraints, and Dependencies:

	Description	Example
Assumptions	List any assumptions made while defining the problem/opportunity.	The willingness of healthcare providers to adopt new technology. Ability to transition healthcare facilities to our new platform before new legislation becomes active.

Constraints	List any known limitations or restrictions.	We are limited in our ability to reach decision-makers at each healthcare facility using our existing network. We do not yet have the right CMO on our team to lead the marketing effort or the right industry connections that we really need. We do not have the funding we need for developing this product.
Dependencies	List any dependent factors or conditions that must be considered.	For healthcare facilities to switch platforms, it will require many levels of approval within each organization and this could take years. We need to be able to expedite those decisions. We will need to support all the existing integrations those platforms support. We will need sufficient capital for this.

Once you have defined the problem and understand the potential constraints and assumptions around it, it is time to turn it into an opportunity statement you can use to get potential investors, customers, and partners excited about your new solution.

TIP

"Most of the time your communication will seem clear and obvious to you but very confusing to others. This is often because you are too close to the topic to see why it's not clear."

After completing more research, you should be able to craft a more refined statement. Here is an example for the healthcare tech startup:

> *Improving electronic healthcare records management in the United States will provide better care, reduce costs, and improve the overall patient and provider experience. There is a niche underserved opportunity at hospitals in the Midwest. There is a market potential of $10 million in annual revenue. Recent healthcare legislation changes give us a window of 18 months to get this to market.*

Without a clear opportunity statement, investors will wonder, **"Why now?"** or **"So what?"**

A clear opportunity statement can help investors, customers, and partners understand your vision while helping you define your product idea.

TIP

"Investors review hundreds of startup proposals each year. They are often looking for a reason to NOT invest and to quickly move on to the next pitch. Give them compelling reasons to consider your idea."

Below is a worksheet to help draft your problem definition. Do not worry about perfecting all this information. In the next stage, the Validation Stage, we will use market research to validate these assumptions. Grab a copy of this template, from the link in the "Bonus Tools and Content" page.

Problem Definition Worksheet

Executive Summary	Your Notes
Brief Description One or two sentences describing the problem/ opportunity.	
Value Proposition *How does solving this problem create value? What kind of value does it create?* (Saves time, eliminates multiple tools, saves money, etc.)	

Problem/Opportunity Statement	Your Notes
Definition Detailed description of the problem/opportunity.	
Need for Solution *Why does this problem require a solution?*	
Opportunity Impact *What is the potential impact of solving this problem?*	

Market Context	Your Notes
Target Audience *Who is primarily affected by this problem?* This might be consumers, an industry, or multiple parties.	
Size Assumption Overview of the market size and segment you are targeting.	
Current Gap *What current market needs are unmet?*	

Importance and Urgency	Your Notes
Relevance *Why is solving this problem relevant at this time?*	
Benefits *What are the potential benefits of solving the problem or seizing the opportunity?*	
Urgency *How urgent is it to address this problem/opportunity?*	

Assumptions and Constraints	Your Notes
Assumptions List any assumptions made while defining the problem/opportunity.	
Constraints List any known limitations or restrictions.	

Dependencies	Your Notes
Dependencies List any dependent factors or conditions, including a minimum amount of external funding, necessary government approval, specific industry expertise, landing an exclusive partnership, business licensing, specific deadlines, etc.	

Other Notes	

Defining a Product Idea

Your product idea uses your problem statement and opportunity statement to craft a business solution.

With a focus on **how** and **why** your business solution is needed in today's market, what makes your idea more compelling than your competitors? How will your product benefit a potential customer? Why will people pay for your solution?

A compelling product idea will help create your brand narrative, find your customer base, and connect with investing partners.

🎓 LESSONS FROM REAL SITUATIONS

We get pitched by hundreds of startups each year and it is often unclear if that team is able to execute effectively to make the business a success. They do not provide any clues as to why they are the right team to make it and how they plan to successfully execute on the idea. They do not provide insights into how they are going to gain market share.

Often they will explain how bad the main competitor's products are and how they could build better products with better features. They believe building new features will allow them to take market share from those competitors. They do not realize how entrenched the competitors are, the integrations and relationships the competitors have, and how much authority the competitor's brand commands in the marketplace.

Consider the following when defining your product idea:

- What are unique aspects of your product idea?
- What target market or customer do you plan to sell to?
- How do your product's benefits compare to your competitors?

As you outline your product idea, it is important to perform some initial research. Consider similar products on the market, how those businesses are structured, and what will set your business opportunity apart in the eyes of potential customers. In other words, what about your idea would compel them to invest time and money in evaluating and switching to your solution? Is it more economical for a potential customer to pay for your solution, switch to your solution, fix the problem themselves, or just keep dealing with the problem?

Defining a Target Audience: Getting A Clear Picture of Your Market

A target audience is the specific group of customers your product serves. They might be businesses or individual consumers, depending on your product and niche. Knowing and understanding your target audience is key to defining your niche market and building a loyal customer base. It also helps you build your brand narrative, appealing to customers with images, taglines, and copy content that connects with them.

To define your target audience, return to your problem and opportunity statements.

Who is primarily impacted by the problem? In the healthcare example, patients, medical facilities, medical staff, and insurance companies are all impacted by inefficient electronic health records. That means each of these may be a potential target customer.

After identifying who is impacted by the problem, you can begin to estimate the size and scope of your target market. From here, narrow your focus with market research to refine and validate your target market.

🎓 LESSONS FROM REAL SITUATIONS

Every year, we review hundreds of pitch deck presentations from first-time founders who believe their startup has the potential to be a billion dollar company. They believe they will capture a significant share of the market from entrenched competitors. They often have not spent enough time researching a niche area within that broader market on which to initially focus. They believe they know what customers want and they underestimate the cost of acquiring market share. This becomes their blindspot.

There are many approaches to market research depending on your product and industry. This market research can include:

Surveys

Below are examples of survey questions you could include. These surveys can be performed in person, on the phone, via email, or via online survey links.

- *What is the biggest challenge you face in your daily work?*
- *Would you use our product to help you overcome that challenge?*
- *How much would you pay us each month for this product?*
- *What is the most important aspect you see in our solution?*

- *Why would you switch to this product?*
- *What do you like about similar products you use?*
- *What do you think our competitors are doing better than us?*
- *What do you like about our solution?*
- *What do you dislike about our solution?*

Competitive Analysis

Reviewing competitor products and comparing product offerings can help you assess what advantage you have over those competitors (if any). Look at what types of customers they have, their sales and marketing strategies, their brand communication, how they build product awareness, and what sets their company apart.

> **TIP**
>
> "We do not recommend building products that have no clear competitive advantages."

Focus Groups

Gather a group of 5-10 people online or in person and interview them with open-ended questions to learn about their needs, what they think about the idea, and suggestions they have to improve the idea. You can also show them design prototypes of the product to get their feedback. Work with a facilitator or moderate a focus group yourself to save on resources.

Interviews

One-on-one interviews provide similar feedback to focus groups. They can be performed in person or online, increasing your potential research scope.

Using the healthcare startup example, you might consider how many healthcare facilities are in each city, their annual IT spend on healthcare systems, the growth of facilities in the target geographic area, etc. After this, consider current gaps in the market. *Are there current solutions to this problem? How do they differ from your solution? How does your product address the market gap, unlike your competitors?*

Now, apply similar questions to your research to get a clear picture of your target market. Through specific, targeted market research, you will be able to better understand your problem statement, refine your solution, and hone your product idea.

🎓 LESSONS FROM REAL SITUATIONS

For one of our portfolio companies, we subscribed to an audience research platform. We spent approximately $100 per month on the subscription and discovered many valuable insights about our target audience that we were previously unaware of. For a few hundred dollars, we saved 10s of thousands of dollars in product development efforts had we gone in the wrong direction.

Keep in mind this market research can occur before you spend any money writing a single line of code. But it takes time and for most people it is not as fun as building the product. But this step helps you understand your product, brand, and customer before you invest time and money going in the wrong direction and building a product nobody wants to buy.

Deciding on a Business Model

Your business model addresses how your company profits from your idea. During the Idea Stage, consider what type of business model best serves your product, market, and eventual exit strategy.

🎓 LESSONS FROM REAL SITUATIONS

In establishing one of our portfolio companies, we positioned it upfront as an acquisition target for Microsoft. We considered our initial business model and exit strategy at the same time. As part of this approach, we aligned our technology to Microsoft technology to be a more attractive target for future integration with their existing platforms. We also knew our product would be more attractive as a B2B (Business-To-Business) SaaS (Software-as-a-Service) business model.

While your business model should focus on the more immediate future and get refined as you learn more from the market, it can also be helpful to consider some long-term goals. We will get more into exit strategies in the Scale Stage of this book.

Consider the following questions to build a first draft of your business model, while keeping in mind the research you did earlier about your target market:

- **Product Characteristics:** How does your product get distributed?
  Do people purchase it online as a SaaS product?
  Do they download it from your site?
  Does it require shipping a physical device to your customer?

- **Competitive Landscape:** How are your competitors operating? If existing companies use a specific monetization strategy, this might influence your subscription and/or pricing structure. For example, if they pay for competitor products through monthly subscriptions, they may resist shifting to an annual fee product. Or you might find a gap in the market where customers would prefer an annual subscription fee.

- **Financial Viability:** Is this model affordable?
  The business model needs to be financially feasible. Building subscription businesses requires upfront costs, time, and niche skills. It also takes time to build enough recurring revenue. It can often take many years for a SaaS company to even get to $1 million in annual recurring revenue.

> **TIP**
>
> "Why founders fail: Most founders greatly underestimate the effort required to acquire customers and build substantial revenue that gets them even to breakeven."

- **Target Market:** Who really are your customers?
  Whether you are selling directly to consumers or businesses or indirectly via channel partnerships, your business model will need to reflect this approach.

Some common business models include:

- **Freemium business model:** Users access a limited version of the platform for free. They can upgrade to access premium features.
- **SaaS (Software-as-a-Service) business model:** Users pay a recurring monthly or annual fee to access the platform.
- **Ad-Based business model:** Users do not pay a fee. Revenue is generated from advertising.
- **Transactional business model:** Users are charged based on the number and type of transactions they perform.
- **B2C (Business-to-Consumer) business model:** Selling products directly to consumers.
- **B2B (Business-to-Business) business model:** Selling products directly to businesses, not to consumers.

Your product can utilize multiple business models at the same time depending on your product and market. For example, the healthcare SaaS startup would be a B2B SaaS model in which healthcare providers pay a monthly fee.

"Recurring revenue will be more highly-valued by investors, and recurring revenue will provide more predictable long-term cash flow for operations."

It is important to consider what type of business model matches your skills and experience. For example, as tech investors, we generally focus on B2B SaaS as far as our investment criteria. We find it easier to build companies that sell to other businesses, and we prefer predictable subscription revenue. When we have the option to do annual subscriptions, we prefer that over monthly, since it helps with cash flow and to finance scaling the business.

Since this is the Idea Stage, know your business model could change. It is perfectly okay to refine your business model as you learn from market research and feedback and as real customers use your product.

Prepare for Challenges

Even though you are at the Idea Stage, it is helpful to be aware of common challenges every entrepreneur may face at different stages of their business.

- **Client Acquisition:** Figuring out who your target customer is can be difficult in the beginning. Figuring out who is willing to pay for your product is even more difficult.

- **Lack of Traction:** Lack of traction can be because potential customers do not understand your product messaging or you may realize there is no actual market for your product.

- **Lack of Capital:** This can occur at any stage of the process, especially during periods of high growth.

- **Competition:** Competitors are always a risk. Building brand loyalty and a compelling story while finding your product-market niche can help keep competition at bay.

- **Talent Acquisition:** Having the right leadership in place is key to success, however, hiring the best people early in the startup process is not always possible. A lack of capital and brand recognition can make it difficult to convince talented and experienced people to join your team.

- **Regulatory Hurdles:** Regulations can slow the process, invalidate your business model, or require you to pivot.

- **Innovation Problems:** Staying current, following market trends, and responding to client feedback can help you avoid stagnation. Innovation is expensive, and requires enough capital to fund product development efforts.

- **Scaling the Business:** Scaling can be one of the biggest challenges for many founders. There are risks of not being able to scale fast enough and there are risks of scaling too quickly.

What Not to Do During the Idea Stage

While passion is essential during the Idea Stage, it can often lead founders to skip ahead in the startup process, lose focus, and make decisions without enough research.

The top five most common mistakes we see founders make at the Idea stage of their startup include:

- **Lack of Focus:** Trying to do too much too soon can cause startups to lose focus. It is important to have a narrow focus initially.

- **Too Vague:** Many startups cannot define the problem or solution well, making it difficult to connect to investors and customers.

- **Invalid Feedback:** Founders often ask friends and family for feedback on their ideas instead of asking for feedback from their target audience.

- **Seeking Investment:** Seeking investment too early can burn contacts and potentially give away too much equity.

- **Ineffective Team Building:** Building a team too quickly or creating a team with the wrong skill sets and experience level can waste resources. At this point, you have not yet refined your idea enough to know the right talent you need to pull this off. Building a team prematurely can waste time and money.

- **False Competitive Advantage:** Many founders believe they have a competitive advantage when they do not. For example, recent AI innovations tempt founders to build entire businesses around a cool new feature they implemented with AI, believing they will take business from entrenched competitors. More often than not, that feature can and will be developed by an existing competitor who has greater resources than you.

While these mistakes are common, it is important to remember this is only the first stage in your startup journey. We recommend setting aside one month for the Idea Stage. This will help you avoid common mistakes and build a strong foundation for a successful business, without wasting resources. We do not recommend spending more time on the Idea Stage because you will continue developing and honing your idea in later stages.

Idea Stage Takeaways

Before you create your startup, the first step is to pinpoint the problem you intend to solve.

- The Idea Stage is the first stage of a startup, where people come up with solutions to problems in their areas of expertise. This stage focuses on a problem, the proposed solution, and how that solution is of value to a customer.

- Every business starts with an idea. Throughout the lifecycle of a startup, the idea can change and evolve.

- Many startups cannot define the problem or solution well, causing them many problems down the line.

- Do not spend too much time at this stage.

- Do not assume your target audience will immediately understand your product and messaging.

Once you have completed the steps above and can clearly articulate your problem and solution in concise, easy-to-understand language, it is time to move to the Validation Stage.

Your Notes

Do Not Pass This Stage Until You

- ☐ Define your problem
- ☐ Define your product idea
- ☐ Define your target audience and niche
- ☐ Define your business model
- ☐ Have confidence your idea will bring value to potential customers
- ☐ Believe it is feasible to implement the solution
- ☐ Can clearly communicate the problem and solution

02

Validation Stage:
DEFINING YOUR MARKET

2 Validation Stage: Defining Your Market

During the Validation Stage, you can refine who your true customer is through market research and real-world feedback.

First, imagine your ideal customer, also called an **ICP (Ideal Customer Profile)**. Then, refine your ICP through interviews with actual potential customers. This market validation is an essential step to narrow your scope and refine your ICP.

For example, suppose you are selling a product to manufacturing companies. You might believe every manufacturing company in the world could be your customer. This is unlikely. During validation, you will find your actual customer profile which should be much more specific, such as "food manufacturing companies based in New York with under 50 employees" as an example.

🎓 LESSONS FROM REAL SITUATIONS

One of our commercial clients was launching a social platform for restaurant venues in the United States. This business had a B2B (Business-to-Business) and B2C (Business-to-Consumer) aspect to it. They were planning to launch in multiple countries and multiple cities within each country.

They decided to first launch in New York City. During this process they learned a lot about the market, made inroads with certain venues, and adapted their platform and customer acquisition process based upon local market feedback.

If they had launched in multiple countries and cities at the same time, they would have exhausted all capital and spread themselves too thin in trying to adapt to each local market.

By narrowing their market focus initially, they could first test and adapt to one market before expanding to others.

Narrowing your customer profile will help you build a brand focused on a more niche audience, saving you a lot of time and money. This will also help you refine your idea, market, and business model for success.

Check the "Bonus Tools and Content" section for recommendations of online tools you can use to help refine your ICP.

TIP

"If your ICP is everyone, keep working at it until you better understand your customer. You cannot sell to everyone. You have limited capital and need to focus. You can always expand your ICP later on."

This stage could take anywhere from a few months to a year, depending on your product and the potential markets. In this stage, you will build on your previous efforts with a focus on:

- Determining your ICP.
- Receiving market feedback.
- Strategizing how to get your product to market.
- Determining the team you need to launch your business.
- Preparing and conducting initial fundraising discussions.

Defining Your Ideal Customer Profile (ICP)

Every business should know their ICP (Ideal Customer Profile).

Depending on your product idea, the ICP might define such things as your customers' age, industry, spending habits, hobbies, or location. Defining your ICP helps you identify the characteristics of your potential customers so you can make better decisions on where to focus your sales and marketing efforts.

For example, returning to the healthcare startup from the Idea Stage chapter, if you were looking at a wide range of patients and providers, you may realize you need to narrow your focus to patients and providers working at "healthcare facilities with less than 50 employees." Narrowing your niche further, your ICP might then become "healthcare facilities with less than 50 employees, that have outpatient services and are located in suburban areas."

The narrower your focus, the easier it will be to validate your business idea with real-world customers. If your ICP was broadly defined during the Idea Stage of your business planning, now is the time to narrow the profile. You can always broaden it in the future, if needed.

Below are examples of criteria to narrow down your ICP. You can keep refining this based on market feedback.

- **Demographics** - Statistical data about characteristics of your ICP
- **Psychographics** - Help you understand the psychological and behavioral traits that influence buying decisions
- **Firmographics** - Similar to demographics but for businesses related to your ICP

Use the **ICP Worksheet** to document your specific ICP. Complete sections relevant to your ICP. *Grab a copy of this template, from the link in the "Bonus Tools and Content" page.*

ICP Worksheet

Demographics	Your Notes
Age Range	
Income Level	
Education Level	
Ethnicity	
Gender	
Marital Status	
Occupation	

Psychographics	Your Notes
Values	
Beliefs	
Lifestyle	
Political Views	
Interests	
Hobbies	
Social Media Group Membership	
Email List Membership	
Social Media Influencers Followed	
News Sites Followed	
Personality Profile	

Firmographics	Your Notes
Geographic Location	
Industry Segment	
Funding Level	
Annual Gross Revenue	
Number of Employees	
Industry	
Ownership Structure	

Technology Currently Used	
Funding Level	
Business Stage	
Decision-Making Process	
Pain Points	
Other Notes	

Your ICP is foundational to your startup's future. Everything from sales and marketing strategies to brand partnerships builds upon the ICP. So, take time to really research and define your initial ICP before trying to validate it. Get a list of suggested tools on the "Bonus Tools and Content" page that can help with this.

Validating Your ICP

Once you have narrowed your scope and refined your Ideal Customer Profile (ICP), it is time to connect with potential customers. Your approach will depend on your target market, current resources, and how you plan to invest time and money connecting with potential customers.

Surveys, focus groups, and interviews are some of the most common ways to validate an ICP. These options allow potential customers to share the range of problems they are dealing with in the current market and discuss what they may be willing to pay for someone to provide a solution for those problems.

When interviewing these ICPs, individually or in focus groups, it is important to:

- Ask open-ended questions and let prospective customers lead the discussion.

- Stay curious. The response may surprise you and test your assumptions. By remaining curious you may even find new opportunities you never considered.

- Identify potential early adopters of your product.

Once initial feedback has been gathered, many founders then make use of wireframes (screen design layouts without graphics) and design prototypes (screen designs with brand colors and finalized graphics) to provide a visual representation of how their product might look. This helps potential customers provide feedback on something more concrete.

🎓 LESSONS FROM REAL SITUATIONS

Below are examples of two mobile application screen wireframes from a real healthcare technology project. These were used to help with market validation and obtaining user feedback to refine the product.

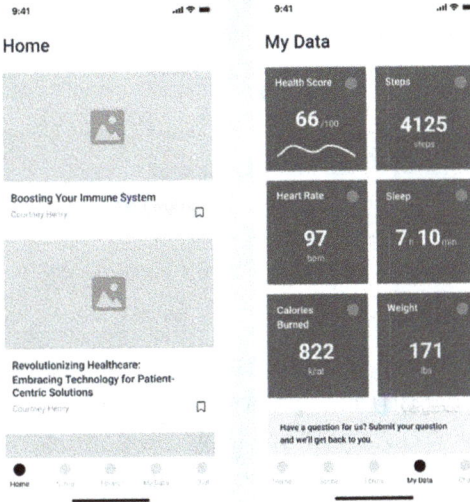

Below are examples of the same mobile application screen in final design form. Even at this stage, these can also be used to get customer feedback prior to implementing them. This will save time and money by helping you to build what customers need.

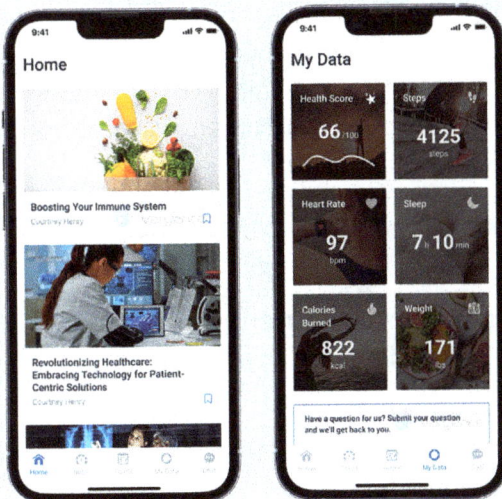

Finalized prototypes can also be turned into clickable prototypes, so that users can select hotspots on the graphics (for example, the Health Score above), and it will then change to show that screen.

Clickable prototypes are a great way to help the user experience the application and get user feedback prior to you investing in development of the product. It is best to **get feedback many times** throughout the design process. This feedback loop will save you a lot of time and money. It will also help you design a product that is more valuable to your ICP.

🎓 **LESSONS FROM REAL SITUATIONS**

When we work with our portfolio companies, our in-house design team creates wireframes and prototypes to help with getting feedback from their ICP.

We share these visualizations in live sessions or let users try them out on their own. Users can click on any part of the screen and leave a comment about where they are confused and add suggestions or questions. Our design team uses this feedback to further refine prototypes.

We often also incorporate these into a video with verbal commentary as another tool to get feedback from users.

Your approach for validating your ICP will depend on various factors, including the industry, the location of your potential customers, and your resources. Most of your initial assumptions will be incorrect. Talking to real potential customers will help you determine which assumptions are accurate.

TIP

"Many founders do not want to challenge or validate their assumptions. They believe their assumptions are already correct."

Refining Your ICP

As you gather more information about your customers, the market, and your niche, you can refine your ICP further.

Just like the Idea Stage, this part of the Validation Stage takes time. Personal bias, assumptions, and emotional ties to your idea can get in your way during this stage. Very often, people will give positive feedback to be nice; it is important to dig deeper and get raw, honest feedback you can use.

When creating an ICP, be aware of some of the below common mistakes that can hinder your progress.

- **Confusing ICPs and Buyer Personas:** ICP and buyer personas are not the same things. Buyer personas focus on an individual buyer (for example, the CFO of a healthcare facility), while an ICP focuses on the characteristics of a company (for example, a healthcare facility with less than 50 employees) that would purchase the product.

- **Targeting too broadly:** One of the biggest mistakes you can make when defining your ICP is setting too broad of an audience. You can always broaden your ICP later. It may sound counter-intuitive, but a narrow focus is better.

- **Not relying on data:** Your ICP should be based on data, both external and internal, along with qualitative feedback. Investors rely on data when weighing the risks and rewards of startup funding. Data can often tell the most compelling story to investors and future partners.

- **Dismissing useful feedback:** Founders can be passionate and emotionally tied to their ideas, making it challenging for them to accept the feedback they receive. Make sure to carefully consider all feedback you receive from your ICP.

By avoiding these common mistakes, you can create an accurate and effective ICP that will help you target the correct real-world customers for your product.

Connecting with Early Adopters

Once you have refined your ICP, it is time to connect with your first customers. **Early adopters** are real customers and users who fit your ICP and are willing to pay to use your preliminary product. These customers might come from your validation interviews, surveys, or via word of mouth from other early users.

🎓 LESSONS FROM REAL SITUATIONS

For one of our portfolio companies, we had some early adopter customers purchase the product over 4 years ago, when the product was first launched. To this day they are still customers and give great feedback. They also got some great pricing deals since they were early adopters.

In exchange for a better deal, they were willing to live with bugs, wait for the product to mature, and help guide the product roadmap.

Beyond getting a good deal, they were also passionate about the company's success, helping refer other people to the product and giving recommendations to others about it on social media.

These early adopters even helped us stay motivated and excited about the product. It was great to see how our product was helping their companies and solving problems.

When we go to trade shows, and an early adopter comes up to say hi, it is very nice to connect with them and catch up.

Finding ways to connect with early adopters can help your startup thrive with useful feedback and free marketing. Here are some ways to connect with them:

- Set up a subreddit on *reddit.com* for your early adopters to connect, give feedback, and stay informed. Engage with them as they ask questions, suggest enhancements, or report bugs.

- Invite early adopters to weekly calls to learn about new updates and ask questions.

- Send a monthly email newsletter with updates and feedback reminders or promote new features.

- If enough are local to an area, schedule monthly local meetups to connect in person.

It is important to **set expectations** with your early adopters. They will have to deal with bugs, glitches, and growing pains as you build out the platform. Because of this, you may want to **consider providing early adopters incentives** like limited-time free subscriptions, referral discounts, early access to beta features, and exclusive perks within the platform. Help them feel like they are part of something special and they will be more likely to spread the word about your company and help you grow by word of mouth. In the early days, every customer will be crucial to your success.

Early Adopters are a great asset for your business narrative and investor pitches, showing real traction. They can be some of your most loyal long-term customers, share their praises with other potential customers, give you detailed feedback, and become some of your biggest fans. Many people love the opportunity to support a new and innovative startup at the very early stages and see it grow into a bigger company. Because of this, early adopters are also great assets to your marketing team. You will be surprised how much early adopters can help you accelerate and champion your business, refer other customers, and lay the groundwork for a community of like-minded people.

Preparing Your Go-To-Market Strategy

Your **Go-To-Market (GTM) Strategy** helps you track, meet, and execute every step needed to break into the market. This strategy will lay the groundwork for your future business success, so it needs to be carefully designed to achieve your desired results. This plan will help prevent financial losses, ineffective marketing, and off-brand targeting. *Use the following worksheet to begin your Go-To-Market planning. Grab a copy of this template, from the link in the "Bonus Tools and Content" page.*

STEP 1
Identify your audience
- Who is your target market?
- What characteristics does your ICP have?
- How does your product differ from other products on the market?

STEP 2
Validate your product
- Define your problem and solution. How does it connect to your ICP?
- What language appeals to your ICP?
- Brainstorm your product messaging here

STEP 3
Set goals and metrics
- Your goals and metrics should be realistic for your industry, your business model, and your ICP

STEP 4
Develop your marketing plan
When developing your marketing plan, consider:
- Pre-launch activities
- Incentives for early adopters and partners
- Marketing channels
- Post-Launch activities

THE **STARTUP** STAGES

Go-To-Market Worksheet

Identify your audience	Your Notes
Who is your target market? *What characteristics does your ICP have?* *How does your product differ from other products on the market?*	

Validate your product	Your Notes
Define your problem and solution. *How does it connect to your ICP?* *What language appeals to your ICP?* Brainstorm your product messaging here.	

Set goals and metrics	Your Notes
Your goals and metrics should be realistic for your industry, your business model, and your ICP.	

Develop your marketing plan	Your Notes
When developing your marketing plan, consider: • Pre-launch activities • Incentives for early adopters and partners • Marketing channels • Post-launch activities Use data-driven approaches to determine the most effective marketing channels for your business before making larger investments.	

Investors look for carefully planned Go-To-Market strategies. Often, founders have an idea but do not know how to successfully execute in bringing that idea to market and acquiring customers. You need to be able to defend your market approach with market research and show you are the person who can execute this and make it successful.

For example, knowing how your product compares to competitors and how you will approach the market compared to them is one of the first things investors want to know.

The more clearly you communicate the *how* (Go-To-Market Strategy) and *why* it is the best approach, the easier it will be for investors to evaluate the opportunity.

Creating an Investor Presentation

Many founders want to jump from the Idea Stage to drafting their first investor presentation. Luckily, by now, you have refined your idea, defined your ICP, and connected with early adopters. This has helped you get more raw feedback and figure out how to better communicate the value you are providing them. **When your ICP is clearly defined, you can better prepare your investor presentation.** This presentation uses data, including market research, to demonstrate that potential clients are willing to pay for your product. It shows investors how your product solves your ICP's problems.

This data will be used to craft an investor presentation, which is sometimes referred to as a pitch deck or funding deck. This presentation will showcase your business idea, traction, and viability to potential investors. Investors see countless pitches every day, so it is best to avoid wordy slides and lengthy presentations. Help potential investors quickly understand the opportunity without overwhelming them with information. If it is a good match, you can enter into more detailed discussions later on.

Below are the 12 slides we recommend including in your investor presentation.

1. Cover Page

Your cover page is the first element of your brand that potential investors will see. Keep it clean and professional-looking with your business name, logo, and the date of the presentation.

2. Problem

Offer a clear explanation of the problem your product solves, including why this problem exists and current market solutions that aim to address the problem. Use quantifiable data where possible.

3. Solution

Provide a clear and simple explanation of how you plan to solve the problem. Include why solving the problem is important, why it has not been solved before, why this is the right time to solve it, and why you are the one to solve it. Try to include specific figures or data to illustrate the impact, including your industry expertise. Focus on the benefits to your potential customers.

4. Traction

Showing your current progress can build trust with investors. Share progress on market validation, building a following (of potential customers excited about your product), customer acquisition, user engagement, and early adopters of the platform. Traction validates the demand for your product in the market. Depending on the situation, you may want to move the traction slide before the problem slide to pique the interest of the person you are presenting to.

5. Competitive Analysis

Part of your deck should tell investors what makes you different from your competitors. Why are you the customer's first choice? Describe how you will set yourself apart from competitors and how you will build a sustainable, unique, economical advantage. Explain why it will be difficult for others to copy your approach.

6. Your Team

Investors want to invest in a team they believe can make the startup successful and provide a return on their investment. Explain why your team possesses the skills and qualities needed to make this business successful. Highlighting any relevant past achievements or industry expertise that will contribute to success helps build trust with investors.

7. Use of Funds

Communicate how much capital you are trying to raise, investment terms, and your plans for utilizing the funds. Explain what you plan to give in return for the investment. For example, "Raising $500,000 in funding this year in return for 10% equity. 40% will be used for product development, and 60% will be used for customer acquisition activities."

8. Market Potential

Market potential is the realistic market opportunity for your business. The market potential for your business can help determine what type of investors you should consider pitching to. For example, if you are looking to build a business that achieves $1 million in ARR (Annual Recurring Revenue) in 5 years, venture capitalists will not be interested since they are seeking billion-dollar opportunities. When estimating your market potential, provide quantifiable data to illustrate the potential size and value of the market. Make sure you can communicate a thorough customer acquisition plan when asked.

9. Market Validation

Investors need to know your product is in demand. Your market validation includes confirmed demand from early adopters willing to pay for your product. Share evidence that will help validate the need and market potential for your solution. While revenue and Letters of Intent (LOI) to purchase are the best indicators of market validation, other methods of market validation include results from surveys, interviews, competitive analysis, waitlists, and beta testing.

10. Target Market

Describe your ICP based on any completed market research and market validation. You may have one or more ICPs you will be selling your product to, however, it is best to narrow your initial focus as much as possible. These profiles should be based on real discussions with potential or existing customers.

11. Revenue Model

Investors need to know how your business will make money. Explain the business model, such as SaaS, B2B, B2C, monthly or annual subscriptions. Explain any other monetization strategies, such as in-app purchases, advertising, or affiliate marketing. Provide forecasted revenue, expenses, and profits.

12. Contact Info

Your final slide should let investors know how to get in touch with you. Share your contact details including email, LinkedIn URL, mobile telephone, website, and other relevant information.

As you build this presentation, avoid personal bias and assumptions. Build the case for your business with data and traction. Keep the pitch clear and simple without too many details that can bog down the presentation. The more traction and validation you can show, the fewer doubts investors will have about whether the idea will be viable.

TIP

"If you have no traction then you need to sell your vision, the market potential, and your ability as the right team, at the right time, to execute on this vision."

Grab a copy of a professionally designed pitch deck template from the "Bonus Tools and Content" page.

Deciding Which Types of Investors to Approach

It is very common for early-stage founders to spend a lot of time pitching their vision to **Venture Capitalists (VCs)**. VCs are interested in very high-growth, high-potential companies that operate in high-potential market sectors. Before spending time pitching VCs, it is important to realize whether your company is a fit for a VC.

🎓 **LESSONS FROM REAL SITUATIONS**

We made this mistake with one of our portfolio companies a few years ago. We got invited into several invite-only pitch events where you get selected to meet with a few investors per week to pitch your product. Despite a lot of great meetings with interesting people, none of them resulted in funding. The reason was simple; we were not building a billion-dollar company. Even though the business was profitable and we had bootstrapped it to over $1 million in ARR (Annual Recurring Revenue), it was of no interest to a VC because they could not see how to scale it to a company worth a billion-dollar valuation. Not only were they right, but we realized this was not the company's ultimate goal. That pitch time could have been better spent growing the business. This was a great reminder that knowing "who your investor is" can keep you on the right path for your product and business goals.

VCs are not the only source of funding. In fact, we rarely recommend founders consider VC funding. **Below are some of the most common types of external investors from whom you might consider raising capital.** We would recommend initially focusing on the first 4 shown below.

There is nothing wrong with bootstrapping your company and not taking on any external investment. If you are able to do this, this can often be your best option for retaining equity and retaining control of your company and future.

- **Friends and Family:** In the beginning, it is best to raise money from friends and family who believe in you. Be sure to let them know this is a risk and they could lose their investment. And keep in mind you do not want to ruin relationships over this.

- **Seed Investors:** These are individual investors who focus on very early-stage companies even before a product is launched.

- **Angel Investors:** These are similar to seed investors but often are entrepreneurs themselves who provide guidance, help, and support.

- **Family Offices:** These are firms that manage the wealth of high-net-worth individuals. They are a great new source of investment that many startups have never heard of and do not know to consider. Today, family offices are increasingly interested in startup ventures.

- **Corporate Investment Groups:** These are groups within companies that often focus on niche areas aligned to their company. For example, a pharmaceutical company may have an investment arm for pharmaceutical startups.

- **Crowdfunding:** This mode is increasingly common through online platforms, leveraging individuals who want to invest or donate funds to help you get launched. There are often rewards, products, and other incentives for crowdfunding investors. Be aware this also takes a lot of effort and preparation to conduct a successful crowdfunding operation.

- **Venture Capitalists:** These are firms focused on creating very big companies in very short timelines. They focus on companies with billion-dollar potential. They have very aggressive requirements for growth and are known for driving companies to "grow at all costs," even to their detriment, as they often have portfolios of hundreds of companies, knowing some will survive and some will die.

- **Private Equity:** These investment groups invest in more mature, private companies. They are often actively involved in the business at the board of directors or management level.

There are many other sources of funding you could consider that will not be touched on in this book, such as **grants from governments and non-profit organizations.**

Keep these options in mind if you are looking for external funding, but keep in mind there is nothing wrong with bootstrapping.

How to Approach Investors

Qualify investors

Before approaching investors, understand what type of investor to approach. See the prior chapter on Deciding Which Types of Investors to Approach.

Create relationships before you need them

Almost every founder I interact with has waited till they need to fundraise in order to build relationships with investors. Similar to general networking, building relationships takes a lot of time, and it is best to build them before you need them.

Pique interest

When reaching out to an investor, do not overwhelm them with a lot of content. Keep it short enough to pique their interest in getting to the very next step, which generally will be an introductory call.

Lead with traction

If you have traction, make sure to talk about it.

🎓 LESSONS FROM REAL SITUATIONS

I have been in many pitch meetings where the investor is looking for a reason to say no and pass on the opportunity.

Once the founder starts to communicate traction, the investor takes the discussion much more seriously because this demonstrates that the founder is able to execute and solve real problems that customers are willing to pay for.

Below are examples of information that is very helpful for an investor to be aware of:

- I have 200 potential customers who have committed to signup and pay $100 per month as soon as you launch. I need funding to get this platform in their hands.
- I already have revenue of $900 ARR per month. I need funding to scale this to $10K ARR per month within 1 year.
- I have already launched the MVP and signed up 10,000 customers. I need funding to scale this to a version 1 and onboard the next 10,000 customers.

TIP

"If you have no sales, it is easy for an investor to argue someone might not be interested in your product. Getting some initial sales helps remove doubt."

Test and refine

Go slowly, try out your pitch and your wording with a few people at a time. Keep refining it as you see what works. The results and approach will vary based upon how you are communicating it (email, chat, text, verbal, in person), your relationship with that person (cold outreach, referral, close relationship), and how strong your content is (your traction, revenue, etc).

Once you feel you have nailed the approach and messaging, then scale it up to more contacts at a time.

Look professional

I receive hundreds of pitch decks per year. Most of them look awful. It is not to say design is everything; the content is most important. But I do notice that when someone submits a professionally designed investor presentation, it stands out. When you are reaching out, you will get better results if you have a professionally designed pre-launch marketing website and investor presentation.

Below are examples of a pre-launch marketing website and investor presentation we did for one of our portfolio companies.

Grab a copy of a professionally designed pitch deck template from the "Bonus Tools and Content" page.

Note: Names and confidential data redacted

Defining Your Initial Organizational Structure

Your initial leadership organizational structure, like your initial idea, will change over time. Consider who you may need to help launch an MVP, get funding, and scale the business. As you progress through each startup stage, your team will scale with you. It is important to have an initial leadership team to show investors you have the industry expertise and the ability to manage finances, acquire customers, and scale. Very often, this is at least a team with a CEO and CTO (technical cofounder). However, depending on your business, you might also have a CFO and CMO.

TIP

"A great CFO can be invaluable to a startup. First-time founders often do not realize the value a CFO can bring even in the early stages."

While cost and productivity are important, company culture fit is also important during the early stages of a startup when people are likely to work long hours, go through the ups and downs of a startup, and work under intense deadlines. The startup culture is not for everyone, so you need to consider the right team members who will fit in this work environment. You may also consider outsourcing for certain skill sets you do not have on your team. For example, UX design, marketing research, cybersecurity, and product development.

🎓 LESSONS FROM REAL SITUATIONS

One of the most valuable lessons I have learned is that having the right people on your team is most important. With the right team, you can adapt and persist until you are successful. With the wrong team, even with an amazing idea, you have very little chance of success. The people you partner with matter more than the initial business concept.

I have never seen an initial startup that did not pivot and refine its direction. The initial business idea looked very different from where it was 6 months or 1 year later. What didn't change was the core leadership team.

For the successful companies, the core leadership team was in it for the long haul, able to adapt and work together to refine the business based on market feedback. For many unsuccessful companies, the founders could not handle pivoting, endure the ups and downs, or adapt to changing circumstances.

Use the Organizational Structure Worksheet below to plan your team composition. This will also help you show an investor how you have thought through the use of funds at each stage.

Organizational Structure Worksheet

Below is a worksheet you can use to model your initial and future organizational structure. Customize this to your specific situation, stage, and budget. Keep in mind most of these positions can initially be fractional/part-time or as-needed roles. Not all are needed in the beginning, but you should continue to refine this as you get through each stage so you have a plan for how you will grow and scale the organizational structure as your startup scales.

Grab a copy of this template, from the link in the "Bonus Tools and Content" page.

Role	Guidance	Your Notes
CMO (Chief Marketing Officer)	It is generally best to have a CMO on your initial team unless you possess these skills as the CEO/founder.	
1. Digital Marketing Manager		
2. SEO Specialist		
3. Content Marketer		
4. Public Relations Manager		
CTO (Chief Technology Officer)	It is generally best to have at least a CTO/technical cofounder on your initial team to help guide technical decisions from the start. This can be an individual or a company that has the right resources to assist in this capacity.	

1. Developer		
2. QA		
3. Security Engineer		
4. Project Manager		
5. Solution Architect		
6. DevOps Engineer		
7. Business Analyst		
8. Designer		
CFO (Chief Financial Officer)	Can initially be a fractional CFO to help with financial modeling. There will not be enough work for a full-time CFO at this point.	
1. Accountant	Generally needed once you get to the Market Fit Stage	
2. Financial Analyst	Generally needed once you get to the Market Focus Stage	
CRO (Chief Revenue Officer)	Generally not part of the initial founding team.	
1. Business Development		
2. Account Executive		
3. Customer Service Manager		
Other Roles		
1. Industry SME (Subject Matter Expert)	This will often be the CEO/founder.	

2. Product Manager	Generally needed once you get to the Market Fit Stage.	

Creating a Financial Model

Managing finances is important for startups. Creating your financial model is the first step. A financial model is a numerical reflection of your business model. It shows investors the relationship between product costs, executive costs, and income needed for financial sustainability.

When investors evaluate an opportunity, they might ask:

- *How big is this potential opportunity?*
- *What are revenue growth projections?*
- *What will be my return on investment and how long will it take to get?*
- *What are the fixed or variable costs as this scales?*
- *Does the founder have a plan on how to manage finances and navigate growth?*
- *How long will it take to break even?*

To answer these questions, people expect to see well-thought-out financial projections in an easy-to-understand financial model. At this stage of your startup, your model is likely aspirational numbers. If you don't yet have revenue, you need to sell the vision, market potential, and your ability to successfully execute.

Below is a basic example of a financial model that helps paint a picture of the potential income and expenses over the next five years. The top section showing revenue, expenses, and net profit is what you can show in a pitch deck.

Once you find an interested investor, you can use the underlying assumptions for a more detailed discussion.

Grab a copy of this Excel template, from the link in the "Bonus Tools and Content" page.

Pro Forma Income Statement

	YEAR 1	YEAR 2	YEAR 3	YEAR 4	YEAR 5
Revenue	$0	$240,000	$360,000	$720,000	$1,500,000
Expenses	$90,150	$119,880	$160,800	$217,520	$276,310
Net Profit	($90,150)	$120,120	$199,200	$502,480	$1,223,690
Underlying Assumptions					
Per Customer Acquisition Cost	$5	$2	$1	$0.90	$0.74
Variable Cost Per User	$10	$9	$8	$8	$7
Fixed Costs	$90,000	$110,000	$150,000	$200,000	$250,000
Monthly Subscription Price Per User	$50	$50	$50	$50	$50
Qty of Paying Monthly Users	0	400	600	1,200	2,500
Qty of Free Monthly Users	10	500	700	900	1,100
Calculated values used in pro forma income statement tab					
Total Users	10	900	1,300	2,100	3,600
Total New Users	10	890	400	800	1,500
Total Acquisition Costs	$50	$1,780	$400	$720	$1,110
Total Variable Costs	$100	$8,100	$10,400	$16,800	$25,200
Total Expences	$90,150	$119,880	$160,800	$217,520	$276,310
Total Revenue	–	$240,000	$360,000	$720,000	$1,500,000
Total Profit	($90,150)	$120,120	$199,200	$502,480	$1.223,690

At this stage, you will likely need a fractional CFO (Chief Financial Officer) to assist with financial projections and guidance. Investors need to know you have thought through revenue, expenses, and underlying assumptions and can defend your figures. A CFO can increase trust with investors.

It is also important to be realistic about the financial potential based on your past experience building companies, your leadership team's capabilities, the potential within your market space, and your ability to acquire paying customers. If you are a first-time founder, it is unlikely you will be able to convince an investor you can achieve even $10 million in annual revenue within a 5-year timeframe, let alone promise your idea will be the next billion-dollar unicorn. Exaggerating what you can truly accomplish is only going to waste time for you and potential investors, as well as burn potential relationships.

🎓 LESSONS FROM REAL SITUATIONS

We were working with a CEO who believed their startup was going to be a Unicorn (worth over $1 billion).

The CEO had a successful career as an employee in an organization, but this was the first time he had ever built a business, led a team, validated an idea, or put together a Go-To-Market strategy.

He was genuinely trying to motivate the team and get everyone excited about the potential of their business; however, he ended up having the opposite effect.

Team members started to realize the CEO was not making realistic decisions for their current situation. Investors realized he had no track record of entrepreneurial success or being able to grow a company to be worth even $1m, let alone a company worth $100m. Investors did not take him seriously and the rest of the team also doubted his abilities to execute.

While it is great to have aspiration goals and think big, it is even more important to make decisions toward realistic expectations and build a track record of success.

Investing your own capital, or bootstrapping, is a great alternative to pitching investors and a common approach to financing in the beginning. While raising millions of dollars may seem exciting and give a sense of accomplishment, building a profitable, sustainable business unburdened with debt and aggressive investors is likely to yield higher returns for you in the long run. Bootstrapping can include crowdfunding and investing personal savings, enabling you to retain control and company equity.

This approach will also save you a lot of time trying to pitch to investors. You can use that time to acquire customers and set your own destiny.

> **TIP**
>
> "If you have no past business successes, be realistic with what you are capable of achieving."

Learning about The Law

As an entrepreneur, you will gain some level of knowledge of the law, even when you are starting out and completely overwhelmed with trying to get your new startup off the ground.

Some experienced business people may tell you if you haven't yet been involved in a lawsuit in some form, you have not been in business long enough. Over the years I have learned a lot about the law, unfortunately, the hard way.

If you are not familiar with doing business in the United States, you should be aware that you can be sued for almost anything, even for false or frivolous reasons. An allegation can be made in a legal complaint without requiring proof or accuracy. It can cost you a lot of money to defend yourself, even if you have done nothing wrong. These situations can arise out of situations such as partner disputes, account collection activities, or intellectual property claims. **As an entrepreneur, you need to be prepared for difficult legal situations that can drain you of time, resources, and emotional energy.**

🎓 LESSONS FROM REAL SITUATIONS

Knowing nothing about law when I first started out, I had registered a domain name for a new business and invested quite a lot of time and money on a logo and website for it.

Immediately after launch, I received a Cease and Desist letter from a public company. They said if I didn't give them the domain name or turn the domain name over to a certified partner company, they would file a trademark infringement claim against us.

Having paid many years of annual domain renewal fees and spending money on a logo and site design, it was painful to have to give the domain name over to them.

I now realize before investing time and money into a business brand, I must first consider if there are existing trademarks in the United States and other parts of the world. I consider this step part of the validation stage.

You cannot stop every possible situation; however, you can do certain things to help protect yourself from getting into bad situations so that you can then focus more on building a successful business.

One of the best things you can do is **get advice and guidance from good attorneys as you grow your business**. Be aware that attorneys in the United States are quite specialized, so you will need many types of attorneys. Just like any other professional, there are good and bad ones.

Below are some of the key legal aspects I recommend getting familiar with so that you can better engage with your attorneys....

- Trademark law to help protect your company name, tagline, and logo design.
- Copyright law to help protect your content and code.
- Patent law to help protect your inventions.

- Business Contract law to create solid agreements with customers, vendors, and partners.
- Employment law to help ensure you don't violate employee rights.

This section is a very short overview of very big topics, just to point you in the right direction.

We are not attorneys, so please do not take any advice here as legal advice. We are sharing information based on our experience as entrepreneurs. For any legal questions or issues, it is best to talk to a qualified attorney who can give you professional advice for your situation. If you need suggestions on good attorneys for the New York or Delaware area, feel free to reach out and we may be able to recommend someone.

What Not to Do During the Validation Stage

Like every stage of the startup process, it is easy to make mistakes during product validation. **Here are some of the most common mistakes to avoid:**

- **Letting bias impact choices.** Biased, leading questions can provide the wrong feedback from focus groups. Remember, true feedback serves your ultimate goals more than your own bias ever could.

- **Overinvestment in the quality or user-friendliness of your solution.** You want to strike a balance with the amount of effort you put into materials provided to your focus groups. Much of what you produce will be rebuilt, so there is no need to perfect it. You need to build only enough to get substantive feedback. A large portion of your work will change based on customer and market feedback.

- **Fear of pivoting.** The Validation Stage is still a great time for pivoting based on feedback from your focus groups. This might mean pivoting your idea, niche, or target customer to better align with the feedback you are receiving.

- **Assuming people will get your idea.** Without clear communication, it can be difficult for a potential customer to understand a product or solution. Do not assume your communication is clear or easy to understand. Most of the time, a potential customer is going to have a much different understanding of your product than you would expect.

Assumptions and blind spots can be detrimental at every stage of your startup. Remember that while you might know your potential customer needs your product, convincing them is more difficult and more important than your assumptions.

TIP

"People won't understand your idea like you think they will. You are too close to your idea to communicate it clearly."

Validation Stage Takeaways

The Validation Stage helps you validate your data and prepare for the prelaunch of your MVP. **Key takeaways in the Validation Stage include:**

- You must get useful feedback from your ICP.

- It is important to refine your ICP and consider who the product or service is for and who it is not for.

- Early adopters are your allies for immediate and long-term success.

- Refining your business model and financial model are important steps during this stage.

At this point, you should now have a better understanding of the problem you are trying to solve, how your solution addresses that problem, and how interested people are in paying for your solution. You have gained some early adopters and have an initial Go-To-Market strategy in place, a draft investor presentation, and a draft financial model. It is time to move to the next stage.

Your Notes

Do Not Pass This Stage Until You

- [ ] Define your ICP
- [ ] Validate your ICP
- [ ] Get some early adopters
- [ ] Prepare a Go-To-Market Strategy
- [ ] Generate a financial model
- [ ] Create an initial Investor Presentation
- [ ] Consider your initial team structure
- [ ] Have an understanding of how much people would pay for this solution
- [ ] Potential customers have expressed interest to become early adopters

03

Launch Stage:
NAVIGATING BETWEEN "PERFECT" AND "QUICK"

3 Launch Stage: Navigating between "Perfect" and "Quick"

Launching is an art. Once you have refined your ICP and obtained early adopters through market validation, it is time to start the process of **delivering your MVP (minimum viable product)**.

A well-planned Launch Stage considers **timing, community, and momentum**. Many companies fall into the trap of investing too much time and money in earlier stages of the process or trying to perfect their MVP, causing them to run out of capital before they can even launch. The **goal** in this stage is to deliver a product that is between "good" and "perfect" to build trust and gather early feedback.

In this chapter, we will look at defining and developing your MVP, delivering it to your ICP early adopters to grow your customer and audience base, and leveraging customer feedback along the way.

Define the Scope of Your MVP

🎓 LESSONS FROM REAL SITUATIONS

At one time, we worked with a founder who wanted to achieve the perfect design for their mobile application. Because their ICP was ultra-high-net-worth individuals, they felt the design had to be perfect before launch. After more than three complete design revamps with multiple designers, the founder ran out of capital and never got it launched. A competitor did launch their product.

That competitor's product was not pretty at all; however, they did get their app launched, got traction, and started to build their business. The lesson is you need to consider a realistic scope for your MVP. Your goal is to get market feedback, get early adopters, and get traction. But you will never achieve any of those things if you do not set realistic goals or wait too long to launch your MVP.

Successful MVPs compromise between the product's overall objectives and requirements. This means developing a basic MVP that delivers useful features for early adopters and the opportunity for actionable feedback and improvement. **Perfection is the enemy** when trying to define the scope of an MVP. Consider the bare-bones version of your product that provides value to your ICP and enables you to gather sufficient feedback for future features.

There are some core features most MVPs need for SaaS products. These include:

- Users must be able to sign up and log in.

- The platform must be accessible in some form, whether in a mobile app store or a website.

- Basic security functionality to ensure people trust you will protect their data. This can be further strengthened in the future.

- There may be regulatory compliance considerations, depending on your industry and product.

- Decent, professional-looking design. This can be revamped in future versions, so do not get caught up on this.

- Minimum core functionality to demonstrate how your product solves a problem for your ICP.

Many founders fall into a trap when it comes to core functionality. They are tempted to create a big list of functionality that extends beyond their current resources and timeline. This can lead to overspending without adding any real value to meet your ICP's needs. It is important to remember that an **MVP is a Minimal Viable Product, not a complete and fully-functional product** with all the features you envision. For many startups, scoping out the features to include in their MVP can be the most difficult part of moving into the Launch Stage. Scope creep (continual expansion of the project beyond the initial scope) and uncertainty about what is most important to include within the scope can lead to slow decision-making and stifle progress. Once you deliver your MVP and receive feedback, you can begin adding valuable features and altering your product for your market needs.

Utilizing Early Adopter Feedback

Feedback is key to improving your MVP and preparing for the next stage of your startup. Combining multiple types of feedback, such as surveys, app analytics, and interviews from early adopters, provides the most holistic picture of user trends and overall product success. While someone might not provide completely honest feedback in a focus group, real users in the market will generally not hold back from sharing their thoughts. Feedback can help develop your action plan and avoid costly missteps.

🎓 LESSONS FROM REAL SITUATIONS

When we first launched one of our portfolio company products, it had an iOS and Android mobile app that was quite limited. Most users of the platform used a browser-based interface and not the mobile apps. Most early adopters understood these mobile apps were limited and appreciated they had some level of functionality they could use if needed.

There was one customer in particular who wrote a terrible review on the Google Play store about the Android application. We finally figured out which customer it was linked to and messaged him to offer additional support, guidance, and help answer any questions. He ignored us and never changed the review. It was very frustrating. Not every user is going to be reasonable and give you grace. Some are going to be very hard on you. If you know this ahead of time, it helps a little bit, but it can still be painful.

When receiving feedback, it is important **to prioritize improvements based on effort, cost, and alignment with your larger business plan.** Just like other stages of the startup process, the ability to adapt and better serve real customer needs is crucial during the Launch Stage.

How to respond to early adopter feedback about your platform

01 GATHER FEEDBACK

Utilize various methods such as surveys, interviews, focus groups, user testing, and app analytics. Try to understand the underlying reasons for their feedback.

02 ORGANIZE FEEDBACK

Categorize feedback into themes and common issues. Try to identify patterns.

03 PRIORITIZE FEEDBACK

Evaluate based on estimated business value, customer impact, and alignment to vision.

04 ANALYZE FEEDBACK

Perform initial business and technical analysis on feedback to understand how it may impact other parts of the platform, and the complexity involved.

05 DEVELOP ACTION PLANS

Create plans to incorporate the feedback. Assign ownership and set clear deadlines.

06 COMMUNICATE WITH EARLY ADOPTERS

Keep early adopters informed of progress and approximate timelines for the product roadmap. Demonstrate commitment to responding to their input.

07 IMPLEMENT CHANGES

Apply changes to your product and monitor how it impacts user engagement and usability.

08 ITERATE AND IMPROVE

Continue gathering regular feedback to improve the product.

Get more templates and take a free startup quiz on 6StartupStages.com

STARTUP STAGES

Define Sales and Marketing Strategies

It is important to have a strategy for how you will manage sales and marketing prior to launching your product, during the launch, and post-launch.

Pre-launch strategies are often overlooked. The pre-launch is your chance to get people talking about your product before it even launches and build a waiting list of potential early adopters. Most of the time founders will launch their product and then start an awareness campaign to get people to use the product. Imagine instead launching your product when you know you have an entire community waiting to use it that is excited to help you succeed. This is what a pre-launch strategy is all about.

TIP

"If you are launching your product and no one has ever heard of it, you missed out on an opportunity. You need to pre-launch your product before the launch."

Consider how to generate a groundswell of interest in your product before it hits the market. For example, **pre-launch websites, waiting lists, and early access, invite-only tactics** create excitement and anticipation for your launch. They also allow you the opportunity to provide updates and create brand awareness without extensive investment. Pre-launch marketing can also help with investor discussions, demonstrating market demand and traction for your product.

Enlist the help of experienced sales and marketing leadership to craft an effective pre-launch marketing strategy. Be careful not to craft a strategy beyond your ability to execute in terms of time and resources.

Efforts spent on your pre-launch strategy will also help you define your **launch plan, sales strategy, and marketing strategy** in future steps. When defining sales and marketing strategies, it is important to think of it like your MVP; you do not want to over-invest or exhaust resources on marketing strategies that have not proven themselves effective for your brand. Continue to study and refine your marketing efforts to find the most sustainable marketing tactics for your business and niche focus.

Use the worksheet below to prepare a pre-launch plan. Grab a copy of this template from the link in the "Bonus Tools and Content" page.

Pre-Launch Planning Worksheet

Use this worksheet to help create your pre-launch plan. Remember to focus on your ICP, leverage early adopters, and consider how to best invest your resources.

Pre-Launch Task	Your Notes
Get a pre-launch marketing website live so potential customers can join a waiting list or sign up for early access.	
Consider partners, early adopters, and potential new customers on this pre-launch timeline. Use this time to convert potential customers to early adopters.	
Consider how to get new people excited about your MVP features and launch and how to keep your current early adopters excited and engaged for the release.	
Consider partner promotions and cross-promotional opportunities.	
Engage early adopters with discounts, promotions, referral incentives, and individualized treatment.	

Consider how to leverage marketing tools such as pre-launch websites, waiting lists, surveys, referral programs, email newsletters, and social media.	
Consider how to leverage Public Relations (PR) activities to get speaking slots, get your articles published, and get mentioned in public communications to raise brand awareness.	
Consider how to leverage influencers to get the message out to a wider audience.	

Build Your Initial Product Development Team

Founders often bring in team members from Fortune 500 companies who have zero startup experience. A founder might think an individual's experience working at a name-brand company is more important than a startup track record. Unfortunately, that is not the case. Fortune 500 experience does not necessarily translate to startup culture where you face limited resources, no infrastructure, no existing processes, no organizational support to help you get things done. Founders often realize too late that enterprise experience rarely translates to startup environment life.

Because your **MVP needs to strike a balance between being "perfect" and "quick"**, you will want to rely on the right development team that is a good match for your budget and company culture. If this is your first attempt at developing a software product or launching a startup, you may not be aware of who to hire, what skill sets you need, or how to validate them. For example, before hiring technical team members, consider hiring a **technical cofounder** who guides those hiring decisions. During the early stages of your startup, a smaller team can help produce a good product in a short amount of time. Some of the most important technical team members and technical skills needed at this stage of product development include technical leadership, UX design, development, quality assurance, and project management.

Finding the right team depends on your situation, budget, company culture, and your experience. Keep in mind how a team member's experience will translate to a startup environment. It is important the team members are passionate about your project, dedicated to seeing it through, and skilled in adapting to user feedback. Focus on onboarding team members who will grow and adapt as your company scales. But before you can consider scaling and exponential team growth, you will want to focus on the base team you need to launch and refine your MVP.

Launching Your MVP

Once you have finished your MVP, it is time to launch. Launching your MVP should not be too overwhelming if you have already executed your pre-launch marketing strategy, refined your launch marketing strategy, and generated industry buzz. By the time you reach your MVP launch, potential customers should be engaged, sharing your brand, and genuinely excited to pay to use your solution to solve their problems. **Your MVP launch** is an opportunity to build more momentum, gain more traction, and forge a closer relationship with your early adopters.

TIP

"If you are doing free pilots for customers, you need to question why the customer is not willing to pay for the pilot. If you are bringing genuine value to them and solving their problems, why wouldn't they want to pay for that? Getting them to pay will help finance the business and provide real proof you are solving problems people are willing to pay for."

Before your launch, consider:

- **Feedback:** Plan how you will track user feedback and bug reporting. This could be as simple as a spreadsheet, a basic task management platform, or a more sophisticated software development backlog management tool.

- **System Monitoring:** Set up system monitoring before launch so you can know how the system is performing once it is launched. As more customers utilize the platform, unanticipated performance issues, such as slowness, may appear. These performance issues can frustrate customers, causing them to potentially cancel their subscriptions. It is best to put proactive monitoring in place for performance issues, system crashes, or bugs so you can remediate them before customers encounter them. Monitoring will not capture all issues, but you can continue to refine it over time as your system matures and you scale your product development team.

- **Recovery Plan:** Ensure you have a disaster recovery plan for your codebase, database, and any components of your platform. Even if you are using a major cloud provider, their cloud environment can still fail, so you need to be prepared with a recovery plan. A major outage immediately after launch does not instill trust in users.

- **Onboarding:** Create a plan for user onboarding. You may need to hand-hold new users through the system to familiarize them with the platform. This will take time and resources so you will need to have people ready to support new users.

- **Coordinate Schedules:** Coordinate a detailed launch schedule with all team members to ensure everyone is on the same page. This includes your development, sales, marketing, and customer service teams.

See the "Bonus Tools and Content" page for recommendations of tools you might want to consider for the above items.

Remember that your MVP is supposed to be an immature product. You should expect to encounter bugs, receive a lot of feature requests, and be prepared for unexpected issues to arise. Focusing on customer service, system monitoring, and disaster recovery can help you avoid roadblocks during your launch. Providing amazing customer service to your early adopters can help them be more forgiving of all the issues they will encounter with your MVP.

What Not to Do During the Launch Stage

Some startups give up before they even finish their first MVP. They might get into partner disputes, run out of resources to complete their MVP, or take too long and miss the market opportunity. It is easy to get carried away during pre-launch and MVP development.

The biggest mistakes during the launch stage include:

- Skipping a pre-launch strategy
- Not building a community of early adopters
- Failing to create a sales and marketing strategy
- Trying to perfect the MVP
- Ignoring feedback from customers
- Getting wrapped up in partner disputes
- Hiring bad team members who fail to produce

It is also important not to make assumptions in the launch stage. Below are some of the common assumptions to try to avoid:

Assuming your platform is going to be easy to understand.

Generally, in the beginning stages, you will need to provide a high level of customer service and hand-holding to early customers. What you think is easy to understand may not be easy for others. You have been working on this for many months, but someone seeing it for the first time may not get it without some help. Your communication about your product might not be clear at all. As you mature your product messaging in the future based on user feedback, it will get more intuitive. But for now, it most likely is not as intuitive as you think.

Assuming your platform is scalable.

Your current version might not scale as much as you think. An MVP will not be a mature, full product. You must make sacrifices to get your MVP to the market. As your product matures in the future, it will become more scalable. At this time, you are going to face bugs and scalability issues and run into many unanticipated problems as more users interact with your product.

Often, founders think of scalability in terms of computer technology: does their computer platform have enough storage, computing power, and memory to meet users' demands? Lack of scalability can show up in many forms, including user interfaces.

> For example, if a software has a screen with a drop-down of locations where the user selects the location of their choice, this may make sense for 5-10 locations but if you now have 500 locations in the dropdown, it is no longer a scalable approach.

TIP

"An MVP is not meant to be scalable and you will never be done with scaling. Even to this day, we continue to work on scalability issues for every platform we support. This is an ongoing battle as your business and platform changes."

Assuming you found all bugs before launch.

Users are going to find new bugs you wish you had found. These are going to be bugs that seem obvious now that they have been reported to you but were not obvious during your testing. Sometimes this is because they are using your platform in a way you would never have anticipated.

> For example, we had one company with an end user who reported a bug no one could reproduce. It turned out the user had used the zoom feature on their browser and this threw off the interface and hid a button for them. This was the first time a user used it this way. Our QA department did not previously test for such a scenario. Users will surprise you with things you never considered before.

As your product continues to mature and scale, you will continue to work through bugs and refine the platform. Even today you will notice major platforms that have been around for decades still regularly release updates to fix bugs and performance issues.

At the launch stage, keep an open mind and do not get upset or offended by customer feedback. It is going to be messy, but the good thing is you delivered your product to the market and are getting real market feedback, even if it is painful to hear. If you were to keep perfecting your product and never got it to market, you would not be able to progress your business and you would have wasted time and money focusing on areas your customers would not care about anyway.

Real market feedback wins over MVP perfection every time!

Launch Stage Takeaways

The Launch Stage focuses on the scope of your MVP and getting your MVP into the hands of your target market and early adopters. **The MVP should have a minimal scope of features to help reduce development costs and get it to market more quickly** to start collecting real feedback and keep early adopters engaged. The ability to make reasonably quick decisions and press forward without trying to perfect is key to getting your MVP launched. An MVP is not meant to be a fully-functional product.

🎓 LESSONS FROM REAL SITUATIONS

As we were working with a founder to get their MVP launched, we reminded them that an MVP is not meant to be a full version of their platform and is not meant to be scalable.

They were shocked to hear this. They didn't understand why we built a platform that was not fully scalable and fully functional. For a new founder, this can be a difficult concept because they want to bring the very best, fully functional, fully scalable product to market from the start.

As they started working with early adopters of the platform, they started to realize that some of the functionality they planned for the fully functional version 1 of their product was not necessary. They realized if they had invested product development time into those features, it would have been a waste.

They understood that by getting a minimal product to market, they were able to get the feedback they needed to truly understand the product roadmap that brought the most value to their customers.

This stage also **focuses on pre-launch, launch, and post-launch strategies** to keep customers engaged and excited. Once your MVP is in use, early adopters and new customers will provide feedback. You will need to prioritize feedback and create and implement a plan for feedback integration. When customers see you value their feedback and deliver what they request, this increases the value of your platform, leading to more traction and momentum. During this stage, you will likely hire a core team or agency to help build and market your MVP. Once these pieces are in place, it is time to move on to the Market Fit Stage.

Your Notes

Do Not Pass This Stage Until You

- [ ] Incorporated feedback from early adopters
- [ ] Defined a Sales & Marketing strategy
- [ ] Launched an MVP

04

Market Fit Stage:
FINDING YOUR PLACE IN THE MARKET

4 Market Fit Stage: Finding Your Place in the Market

In the Market Fit stage, you will focus on **Product-Market Fit (PMF)**: finding your best place in the market for maximum growth and profitability.

> ### 🎓 LESSONS FROM REAL SITUATIONS
>
> A few years ago, we realized one of our SaaS portfolio companies achieved PMF.
> We measured the achievement of PMF by
>
> - A win rate of at least 75% within an ICP.
> - An ARR of at least $1 million for that specific product.
>
> We realized whenever the sales team provided a demo to a potential customer matching our ICP, the potential customer would purchase a subscription to the platform at least 75% of the time.
>
> Once we realized PMF had been achieved, we then focused our product development efforts on improvements that would be most valued by that ICP.

In this stage, you will leverage feedback from your MVP users for defining your first official product, or Version 1.0, with an added focus on brand communications and marketing. **For many startups, the Market Fit Stage will take from one month to a year.** This section will help you create a product roadmap, help scale your first official product, and close more deals.

Refine Brand Communications

Clear and compelling brand communication tells your ICP what you do and how you can help them. Your brand communication should provide a consistent message across your public marketing site, investor pitch presentation, brochures, and platform. Typically, founders are too deep in the weeds to realize their brand communication is completely confusing to someone who has never heard of their product. Remember, early adopters are key to word-of-mouth marketing. When someone lands on your marketing website homepage, they need to be able to understand, within seconds, what your company does and how you can help them.

Here are some questions to jumpstart refining your brand communications. They can also help you determine if it is time to hire a brand strategist:

- When someone who has never heard of your product before visits your public marketing website for the first time, can they understand within a few seconds exactly what your company does?

- Is your brand communication consistent across all mediums, including your website, social media, and email newsletters?

- If you need to reach audiences in different regions, do you need to localize the communication? For example, does your website need multilingual capabilities to reach users?

- Do the Calls-To-Action (CTAs) on your site match the next step actions you would like visitors to take?

🎓 LESSONS FROM REAL SITUATIONS

We realized with one of our portfolio companies' public marketing sites that the CTA (Call-To-Action) buttons were too many steps ahead of where the buyer was.

The prior CTA would be "Buy Now" and "Purchase," however, they were not at that stage of their buying journey.

They needed to learn more about the product, determine whether it met their needs, and understand how it worked.

We adjusted the CTAs to "Watch Demo" and "Schedule Call," and conversion rates increased dramatically.

- Is the communication focused on your ICP's needs rather than purely promotional information? For example, does your website speak to the true needs of your potential customers as opposed to talking about product features they may not be interested in?
- Do the visual elements align with your brand personality and values?

When it comes to **refining a brand message**, independent feedback can give a fresh perspective. A certain amount of refinement can be accomplished from user feedback and market research. However, we would additionally **recommend hiring a brand strategist** to help ensure you have clear brand communication that resonates with your ICP. Responding and adapting to early adopter feedback will also help you refine brand communications and help you turn your MVP into your first official product.

🎓 LESSONS FROM REAL SITUATIONS

We hired a brand communication strategist to help us with multiple marketing sites for our portfolio companies. They brought to light things we never considered were confusing users who landed on the marketing site. We would not have recognized our communication gaps without their help.

For example, the marketing sites were very focused on features of the products, however, visitors to the sites did not understand how those features related to their business needs.

Many times founders believe that the more features the product has will convince a potential customer to sign up. Effective communication is more important than more features.

It is an art to put yourself into the mindset of a customer visiting your marketing site who is not very familiar or maybe has never heard of your product.

Scaling Sales and Marketing

It can be difficult to know which exact sales and marketing strategies will work best for your business.

🎓 LESSONS FROM REAL SITUATIONS

For one of our B2B SaaS portfolio companies, we ramped up an outbound sales team with three full-time employees and hired a fractional sales manager to oversee. We spent quite a lot of money on salaries, tools, and training. We also spent a lot of time on this effort.

We realized outbound sales was not the most effective approach because it was highly unlikely we were going to find the potential business customer at the exact time they needed to consider our product. The results from outbound sales were almost a complete loss.

At the same time, we noticed inbound marketing leads had a very high close rate, so we invested more time and money into generating inbound leads.

We tested which marketing tactics worked best and then ramped up those marketing efforts. The results were a 100% increase in subscription revenue.

This does not mean for every business this is the right approach. You must discover what strategies and tactics are most effective for your business, at this point in time. This will change over time as your business changes and the market changes.

In prior stages, you only had an MVP in place. Now that you have your first official Version 1.0 of the product ready for launch, **it is time to scale up sales and marketing.**

Most startups utilize multiple tools for sales and marketing as they begin to scale. **The most common types of tools to consider include:**

- CRM (Customer Relationship Management)
- Email Marketing
- SEM (Search Engine Marketing)
- Sales Intelligence
- Data Analytics
- Data Integration
- Digital Document Signing
- Virtual Meetings
- Email and Telephone Validation
- AI standalone solutions and AI integrated into apps to help with generating content, meeting summaries, in-call sales assistance, etc.

Check the "Bonus Tools and Content" page for suggestions of tools you can use for your business.

Do not assume what works for other companies will work for you, particularly if they are at a different startup stage. Run tests to evaluate which sales and marketing strategies and tactics yield the best results for your business. This can help ensure you have a more complete sales and marketing strategy, along with the operational resources and budget to execute these strategies. During this stage, consider scaling up your sales and marketing teams and maturing the technology tools they utilize.

> **TIP**
>
> "You need to find the marketing tactics that work best for you, where you are currently at in your business. You must test and see what works."

Work with your CMO using the below **Marketing Roadmap Worksheet** to help identify the most effective areas of focus for your startup. This will vary based on your marketing strategy, resources, and timelines. *Grab a copy of this template, from the link in the "Bonus Tools and Content" page.*

Marketing Roadmap Worksheet

Area	Key Examples	Your Notes
On-Page SEO	Improve website title tags, meta descriptions, alt image text, AI overviews, etc.	
Technical SEO	Create sitemaps, improve page speed, and optimize website structure.	
Content Optimization	Improve old content and blog posts.	
User Experience and Conversion Rate Optimization	Audit site design, improve navigation, analyze heatmaps of user behavior.	
Local SEO	Add local schemas for city and address locations, adjust content for local landmarks and events.	
SEO Benchmarking	Track ranking on topics and assess traffic patterns.	
Topic and Keyword Research	Identify new topics and keyword opportunities.	
Analytics and Tracking	Create custom analytics reporting. Use tags and triggers for tracking conversions from end to end.	
Outbound Marketing	Email marketing via direct mail and newsletters.	

Inbound Marketing	SEO, SEM (Search Engine Marketing), and Paid Advertising campaigns.	
Events Marketing	Trade shows, partner events, and hosted events.	
Community Building and Engagement	Forum marketing, industry groups, contests, and referral programs.	
Off-Page SEO	Backlink building, social media marketing, public relations, influencer marketing, and podcasts.	

Scaling Product Development

At this stage, as you receive feedback on your MVP and use that feedback to refine the scope of your first official Version 1.0 product, you will likely need to scale up your product development team. You may need additional skills in your team, such as cybersecurity, automated QA, compliance, and additional senior-level technical resources.

Below are considerations when scaling your development team. Your technical leader should have experience to guide these initiatives alongside the rest of your leadership team:

- Functional management of areas such as QA, Solution Architecture, Development, Business Analysis, and Project Management.
- Update and formalize policies, processes, procedures, and standards to support onboarding new employees and business efficiency.
- Create standards for how the team should develop, test, and deploy updates.
- Adopt agile methodologies that prioritize customers. Agile development methodologies focus on shorter development cycles and provide frequent product updates. Prioritize

product updates that provide the most value to your customers.

- Prepare for necessary security and compliance updates. For example, you might start to target enterprise customers who require a certain ISO certification. Those customers will want to know you operate your business using best practices that safeguard their data. Plan for the efforts and resources necessary to meet these compliance needs.

- Improve team resiliency by identifying mission-critical roles and providing cross-training in case a team member is absent or leaves.

- Incorporate more automated testing into the development process. This includes testing at the programming level and user interface testing to catch issues before consumers report them.

- Incorporate security vulnerability scanning during the development process to address security concerns before the product is released.

- Ensure you have enough senior-level resources in each functional area to lead these efforts. For example, your QA team may need a more experienced manager to lead less experienced QA resources, helping direct their efforts, defining processes and procedures, mentoring, and helping them overcome challenges in their daily work.

- Automate operational and development activities wherever possible.

- Audit software architecture to find areas for improved security, scalability, and resiliency.

- Increase activity logging to understand user interactions to improve the user experience and user interface.

As with every stage of this process, focus on data-driven goals and results when scaling your team. **It is important to have technical leadership helping guide these decisions.**

🎓 LESSONS FROM REAL SITUATIONS

We started helping a health tech company many years ago during their Idea Stage. They started with zero technical resources so we worked together to brainstorm platform features, prototype designs, build the POC (Proof of Concept), and launch their MVP.

Eventually, they were acquired by a public company and had to comply with more industry regulations.

They added US technical leadership and developers to our product development team that continued helping from our Ukraine office.

To comply with regulations surrounding their insurance reimbursement model, only their US team could have access to production-related data. This meant that scaling their business and team resulted in new compliance regulations that dictated their team structure.

In earlier stages, using an outsourced team can help you with the depth and breadth of skills you need to launch your business. It can be difficult to recruit, attract, and retain all the team members you need in the earlier stages because you cannot afford to have them involved on a full-time basis. **At this Market Fit Stage, consider starting to build out an internal, in-country team to complement existing outsourced team resources** and be mindful of which positions you scale.

Creating a Product Roadmap

At this point, you have started to build a backlog of feature requests from customers. Prioritizing feedback during this stage can be overwhelming as you develop new features and build your customer base. You will likely struggle to get a consensus from all team members on an execution plan. A product roadmap can help align all areas of your company and clarify goals.

You can create an initial product roadmap in a spreadsheet. Below is an example to model it after:

- Begin by listing the core areas in the first column.

- Then add additional columns for each future month and one column for areas that are completed and live.

- Then shade the respective month cells to show the timeline for each area. Include the current percentage completed.

As an example, we can interpret from this report that the Patient Health Score section being worked on in Month 1 is now 31% completed, is the main focus in Month 1, will only take one month to be completed, and is targeted to be live in production at the end of Month 1.

Product Roadmap

AREA	COMPLETED & LIVE	MONTH 1	MONTH 2	MONTH 3	MONTH 4	MONTH X
Patient						
1. Health Chart	100%					
2. Health Score		31%				
3. New Patient Report			0%			
Doctor						
1. Add Patient Notes				0%		
2. Adjust Recommendations					0%	
3. New User Interface					0%	

Get free templates and take a startup quiz on 6StartupStages.com

"A product roadmap can align the company. Very often different departments are operating in isolation and not in sync."

Using color coding and marking production progress towards 100% complete makes it easy for others on your team to follow the roadmap.

We use the below color coding with our roadmaps. In Project Management terms, this is called RAG status, which stands for Red, Amber, Green.

- **Green:** Everything is on track. There are no known risks affecting the timeline.
- **Amber:** There are potential risks and the timeline is tentative.
- **Red:** Known issues are affecting the timeline.

Knowing when a new feature is set to launch helps marketing resources determine promotional timelines, sales to inform new or potential customers, and customer service to prepare to support new features. Keeping everyone aligned gets more difficult as your company scales, but even a simple product roadmap can help align the entire company.

🎓 LESSONS FROM REAL SITUATIONS

After one of our B2B SaaS portfolio companies launched their initial MVP, we were trying to finalize the Version 1.0 design.

During that process, we were closing increasingly larger deals and almost every new deal came with additional product feature requests to deliver.

This saddled our development team with a large backlog of tasks. We needed to decide which exact feature requests to put on our product roadmap for Version 1.0 and how we would prioritize them.

At the same time, we had to be careful with commitments we made to potential customers and set realistic timeline expectations.

Even to this day, as we close larger deals, very often there are requests to add items to the roadmap that are specific to those customers.

Finalizing Version 1.0 Scope and Launching

After having launched your MVP, it is important to plan the launch of your first fully functional stable, scalable version, called Version 1.0. The MVP helped you gather market feedback and get early adopters but was not meant to be stable, scalable, or have all the necessary features. As you finalize Version 1.0, incorporate all feedback received since the launch of your MVP and consider new requests required to win new deals.

Balancing demands from current users versus demands from potential new customers, while also managing limited resources, is a struggle you must manage throughout the life of your startup.

As you finalize your Version 1.0 product scope, keep the following in mind:

- Customers demand high-quality craftsmanship that minimizes bugs.
- Users expect easy user experiences that keep them engaged.
- You are expected to safeguard business and customer data. Security is important.
- Focus on technical decisions that get your product to market quickly without jeopardizing your future ability to scale.

- Develop an organized tracking system for feedback to help prioritize future product versions and features.
- Feedback from your ICP is the highest priority.

Every feature request carries an internal cost to get it developed and delivered. In the real example above, we realized some of those customers requesting new features were not ICP matches. If we invested in building out features for customers that did not match our ICP, we risk delaying features that we need to get in place to solve our true customers' problems. There were many situations where potential customers who did not match our ICP would make feature requests, we would build those features, and they would never follow through with their purchases. Those were painful and expensive lessons.

Once you finalize your initial Version 1.0 scope, you need to get it developed and launched. It is essential to adopt an **iterative and agile approach**. Do not wait to complete every feature before launching Version 1.0. Instead, release new updates every 2-3 weeks to pace your development and marketing teams. This will keep customers excited and engaged. Keep an eye on the roadmap of features you planned for Version 1.0 so you can monitor monthly progress towards your full Version 1.0 scope.

While you try to keep the scope of Version 1.0 stable, you will need to commit to new, unanticipated features to close new deals and **get to PMF (Product-Market Fit)**. Try to manage those timelines and commitments so that you can meet expectations and not strain your resources.

TIP

"Many founders get trapped into an endless scope creep that stretches their resources too thin, sabotaging their own efforts to get version 1.0 delivered to the market."

What Not to Do During the Market Fit Stage

As you develop your Version 1.0, it is easy to get distracted from all the user feedback. It may be tempting to respond to feedback outside your market fit. This is one of the most common mistakes during the Market Fit stage.

Other common mistakes during the Market Fit Stage include:

- **Changing an ICP without new research.** Trust your ICP market research and do not pivot if you do not need to. At this stage, it is crucial to only pivot to meet MVP feedback data. Focus on the task at hand and avoid rash decisions based on anecdotal information. If you decide to pivot, you will need to return to market validation efforts.

- **Neglecting good user experiences.** Good user experiences can go a long way in the early stages of a startup. A good user experience can make up for users having to deal with bugs, limited features, and performance issues. Users may often value a better, simple user experience over a product that has more functionality but is too confusing. User expectations and demands are quite high these days; they want simple, intuitive products that do not require a user manual. It is important to have the right user experience and user interface design skills on the team from the start.

- **Neglecting marketing.** Marketing is a broad term and is often overlooked by companies. Founders often believe that just adding new features will sell their product. It is important to scale your marketing efforts. This effort can include community building, content marketing, paid campaigns, SEO, search engine marketing, public relations, and more. Having the right marketing leadership, such as a Chief Marketing Officer (CMO) on your team, is critical to success in this area.

- **Not prioritizing Compliance Efforts.** Sales and Marketing might hit roadblocks as they try to close larger deals with potential customers who have compliance requirements like SOC 2 Type 2 or ISO certifications. Preparing for these certifications ahead of time can help your team close new deals more quickly in the future.

Remember this stage prioritizes market fit, growth, and consumer satisfaction equally. A lack of balance can easily lead to wasted time and money, or lost customers.

Market Fit Stage Takeaways

The Market Fit Stage **focuses on how you get the product to best fit into the market for maximum growth and client satisfaction.**

Like the Launch Stage, the Market Fit Stage relies on a balance between improving the product and meeting customer needs. As your customer base grows, it is important to consider what feedback to prioritize.

This stage helps **refine the company's niche market focus** and customer base while launching its first official product. Finding the right product-market fit can be an art and generally takes much more time than anyone expects. **No matter how good you think a product is, what customers think is what really matters.** The market is the ultimate judge of your product's value.

If you are not seeing a significant level of growth, remain in this stage until you do. This stage will take, ideally, three months to one year. Once your first product is stabilized and there is consistent, substantial subscriber growth and user engagement, you are ready to move to the Market Focus Stage.

Your Notes

Do Not Pass This Stage Until You

- ☐ Scaled Sales and Marketing teams

- ☐ Scaled Product Development team

- ☐ Launched and stabilized version 1.0

- ☐ See steady subscriber growth and user engagement

05

Market Focus Stage :
ACCELERATING GROWTH

5 Market Focus Stage: Accelerating Growth

The Market Focus Stage is about **increased adoption of your product within your existing ICP.**

🎓 LESSONS FROM REAL SITUATIONS

During COVID-19, one of our SaaS portfolio companies increased sales and marketing efforts to their existing ICP to drive increased product usage.

Having achieved PMF just before the pandemic, we focused on expansion within that ICP, which led to a 5x increase in growth.

We also focused on increased resiliency to accommodate the growth and prevent platform outages. Even to this day, we continue to improve resiliency and health monitoring of the platform, trying to anticipate problems that could arise as the company scales.

In this chapter, you will create a growth strategy that emphasizes company health and resiliency to prepare for the future. We will look at **data-driven strategies for acquiring more customers within your ICP** and ways to increase revenue while maintaining profitability. For many companies, this stage can take years to complete and requires obtaining true PMF before accelerated growth is possible.

TIP

"Many startups raise millions of funding prior to reaching PMF and blow it all. Achieving PMF before major fundraising helps you make less expensive mistakes."

Creating a Growth Strategy

A growth strategy is how you plan to increase market share, increase revenue, and grow your business. Growth is essential during the Market Focus stage. **It is important not to lose track of who your target market is.** Rather than focusing on expanding your ICP, focus on growth within your existing ICP.

Each startup's growth strategy is unique. Strategies we recommend include:

- **Product improvements:** This can include new features, new pricing strategies, customer referral or loyalty programs, as well as software and UX improvements.
- **Channel partnerships:** Find partnerships with companies that can resell or refer customers.
- **Integrations:** Find strategic integrations with other platforms to help cross-sell.
- **Move upmarket:** Building relationships with larger accounts
- **Advancing marketing efforts:** PR and community building are essential for growth. This might include connecting with influencers, participating in industry conferences, or adding new value assessment tools to enhance your marketing strategy.

Managing Customer Requests

At this stage, you should have many feature requests, bug reports, and demands from customers. Now is the time to work through that backlog, show customers and the market you are listening to them, and **ship regular updates** to meet their needs. But before you deliver on those requests, you need to improve your process for prioritizing them. For example, hundreds of customers with the same feature request may be more important than a feature request from one customer. You may find some requests require very little effort but make customers very happy, while others require extensive time, support, or financial investment.

When evaluating each request at this stage, we use a simple data-driven scoring system. Each item is given a score from 1-10.

- **Business Value:** how much value this can bring to the business
- **Business Urgency:** how urgent is this to implement for the business
- **Implementation Complexity:** how much effort will this require to implement
- **Maintenance Complexity:** how much effort will this require to maintain
- **Customer Demand:** how important is this to customers
- **Strategic Alignment:** how well does this align to the product roadmap

Some companies implement **public voting platforms** where customers can submit feature requests or vote on them. This helps you get direct, real-time feedback from customers and helps them feel more engaged. Customers can see all feature request priorities and progress, which can reduce customer support requests. Options like this reduce pressure on your team to connect with customers over each request and help customers feel more connected to your community. You should be aware that this approach does expose your valuable product roadmap and market feedback to competitors. This type of data-driven approach takes the guesswork out of the process and avoids wasting time trying to choose which requests to prioritize.

🎓 LESSONS FROM REAL SITUATIONS

For one of our portfolio companies, we build customer-specific product roadmaps for larger customers.

For example, a larger customer may have a list of demands they requested when they signed up so that we could close the deal with them. Those requests could take one or two years of time to complete. Each month we publish a separate product roadmap and meet with them on a monthly basis to discuss new releases and refine the roadmap based upon their needs.

Focusing on Profitability

Now that your first version has launched and you have obtained feedback for your next product versions, it is important to **prioritize profitability and investor ROI (Return On Investment).**

If you do not consider this carefully, you could spend years of your life building a startup and walk away with nothing. This happens often when founders focus on raising so much capital they cannot achieve the growth required to make a realistic return on investment for their investors or for themselves.

At one time, it was very popular for startups to focus on growth rather than profitability. They focused on growth at all costs. Some startups did not even pretend to forecast a profit. Instead, they planned to continue raising lots of funding, grow aggressively, and then get sold or go public before they ever had to be concerned with making a profit. When funding dried up, they were in a bind because, without funding, they were insolvent. In other cases, investors might swoop in and offer them opportunistic terms that take advantage of the founders before they went bankrupt. This is why focusing on profitability and sustainability is important. It allows founders to build businesses they can benefit from long-term. A focus on profitability enables them to maintain control of their company and their future. Later, they can decide to sell the company, get acquired, or, at that point, raise capital on more favorable terms.

One of our preferred financial metrics for helping SaaS startups focus on profitability is called the **Burn Multiple**.

This is a popular term initially coined by a guy named David Sacks. He was inspired by Bessemer's efficiency score formula and the hype ratio.

This Burn Multiple metric helps you realize how much each additional dollar of Annual Recurring Revenue (ARR) costs your company.

Let us assume you spent $3m to add just $1m more in ARR. This is a Burn Multiple of 3 which is bad. **The lower the Burn Multiple the better. Under 1 is great.**

I have seen multiple startups burn through a lot of money on paid ad campaigns only to add a little bit of ARR. They believed they could buy their way into the market.

Regular financial checks and profit planning can help you maintain a profit focus. In most cases, monthly assessments can help you understand your monthly overhead, the cost of each employee, and any fluctuations in your Burn Multiple that might predict financial instability in the future.

We recommend monthly checks, including:

- Official "monthly close" of your financials run by an accountant.
- Review the income statement and balance sheet.
- Review the Burn Multiple.
- Review fixed monthly costs. For example, you might have to pay a minimum cost for hosting your public marketing site regardless of how many customers visit your website.
- Review variable costs for each customer you acquire. Variable costs are costs that increase or decrease based on the number of customers that are gained or lost. For example, let us say a new customer of your platform uploads files and uses 1GB of data storage, there would be a variable cost for this data storage.
- Understand the overhead incurred for each employee you add to the company.

Whether you work with an outside accountant or one on your team, monthly financial checks help you monitor profitability and financial KPIs (Key Performance Indicators).

🎓 LESSONS FROM REAL SITUATIONS

A CFO can be one of the most valuable team members in your company. Most founders do not realize how much accounting and finance help they need to build a successful company.

One of our portfolio companies has an amazing fractional

CFO. Without him, we would not have been able to craft pricing strategies that closed multi-million dollar deals with customers. He helped us model pricing strategies we had never considered and close deals while also maintaining high profitability.

A CFO is important at every stage of your startup. They are important from the very beginning when you are trying to compose a financial model to the very end when you are trying to determine whether to sell or get acquired by another company.

Each startup will shift its focus to profits at a different time. It is important to consider the sustainability of your product and your company, so you do not lose control of your company or your future.

What Not to Do During the Market Focus Stage

It can be hard to avoid getting carried away during the Focus Stage. Some founders spend too much time at this stage focusing on scaling rather than laying the groundwork for long-term financial success.

Here are the top four mistakes to avoid at this stage:

- Do not stray from your current product focus, pivot into new markets, or make rash decisions. Instead, focus on letting data drive your decisions and growth strategies.

- Do not lose sight of your finances. Add more rigor to your financial oversight and keep your pulse on finances and metrics.

- Do not rely on cash injections. Many startups focus too much on fundraising and not enough on profitability and cash flow. This can be to their detriment, raising so much money, they are not able to achieve the growth required to provide a return on investment.

- Do not lose sight of the key people in your company that make your company successful. Keep your key people motivated and incentivized.

Remember, at this stage, PMF and profitability are important. And stability is more important than exponential growth.

TIP

"There are surprisingly many founders who do not value people. It is important to appreciate all your employees and genuinely care about them. They will know if you sincerely care about them and their success. They are important to your future success."

Market Focus Stage Takeaways

The Market Focus Stage balances profitability with proving product-market fit. During this stage, your product will mature, stabilize, and provide enough value to customers to achieve a win rate of at least 75% with your ICP.

This step can take many years, depending on a company's product, industry, and niche market. If your company and product are underperforming, you may not be aligned with your ICP. In this situation, revisit steps in the Validation Stage to perform further market research and validate who your ICP truly is. Once you have met these goals and have seen steady market recognition, it is time to move to the Scale Stage.

Your Notes

Do Not Pass This Stage Until You

- ☐ Defined a business growth strategy
- ☐ Achieved high deal win rates
- ☐ Achieved profitability
- ☐ Gained recognition in the market
- ☐ Matured and further stabilized product

06

Scale Stage:
DELIVERING GROWTH AND SUSTAINABILITY

6 Scale Stage: Delivering Growth and Sustainability

The final stage in The 6 Startup Stages is one of the goals of most founders: **Scaling.**

During the Scale Stage, you have proven yourself as reputable, valuable, and profitable within your market focus.

The most important focus at this stage is *"How will you scale the business?"*

At the same time, you should start to invest more heavily into improving operational processes, procedures, organizational structure, resilience, and sustainability. This includes **managing business risk, reducing organizational and technical debt, and eliminating bottlenecks.**

Many companies believe they can simply scale up the number of employees to get more results. Adding more people following bad processes means you will spend more money without the intended results. This happens very often even in development teams, where companies add more developers but do not adapt or fix the development processes, resulting in limited productivity improvements.

Whether you are looking to sell your business, get acquired, go public, or create a lifestyle business, this chapter will help you focus on strategies for scaling and expanding for long-term company health, growth, and sustainability. This chapter will also discuss expanding your ICP, expanding your market focus, considering exit strategies, valuations, and what not to do when scaling your company.

Expanding your ICP and Market Focus

Staying focused is difficult. During earlier stages, resisting the temptation to alter or expand your ICP, build new products, or enter into new markets is important so you can stay focused. Upon reaching a stable and profitable point in your business, consider expanding your ICP and your market focus.

🎓 LESSONS FROM REAL SITUATIONS

One of our portfolio companies had spent years developing so many products that it did not have enough sales force to sell them all. They could not figure out where to focus development efforts, sales resources, or business direction. They were profitable and kept investing in new product ideas. However, their growth was stunted because they did not focus on the products that brought the most revenue and the most profits.

It was very difficult for them to let go of some of those products because they knew there was some market for each one. After many years, the CEO realized they needed to focus on core products, and the company started to see much better growth.

Now that they have grown, they have more resources to split among multiple products so they are not stretched so thin. They can afford to focus on more areas concurrently.

Expansion efforts should still be based on market research and data, not gut feelings or anecdotal evidence. **Every expansion effort is going to cost you time and money.**

You might also find that in recent months or years, your industry has already evolved and there are new opportunities due to market trends, economic conditions, regulatory changes, or other factors you did not anticipate.

Below are questions to consider when expanding your ICP and market focus:

- *Are there any additional products we can develop to upsell to existing customers?*
- *Are there any partnerships we can leverage to upsell to existing customers?*
- *Are there partnerships we can leverage to widen our reach of customers that fit our ICP?*
- *Have market changes led to new ICP opportunities? If so, what would be the effort (if any) to adapt the products to meet their needs?*
- *Are there any opportunities to integrate into other platforms and cross-sell with those partners?*
- *What are the untapped markets we can sell our existing products into?*
- *Should we unbundle some features within the platform to increase revenue with existing and potential customers?*
- *Can we afford to invest time and money into this?*
- *How much will this impact other areas of our business?*
- *Is it better to remain focused on what is working?*
- *How will we incentivize our sales team in a way that does not impact other products they sell?*

Each new product you develop has costs you may not initially realize. If you want to pursue a new product or ICP, you need to go through the proper stages to vet the idea. You will be tempted to skip those stages, delegate to people who may not be as passionate about the idea as you are, and throw money at the idea to see if it works. We recommend you start with the Idea Stage to better define the problem and solution. Then, continue to the Validation Stage and subsequent stages.

Other soft costs to consider that may not be obvious in the beginning include:

- Maintaining and enhancing the technology so it does not

become obsolete.

- Training sales teams on how to sell it.
- Training support teams on how to support it.
- Training marketing teams on how to market it.
- Updating brand communication to clearly communicate your offerings.
- Incorporating the new product and ICP into business policies, processes, and standards.

It is very common for companies to get into the trap of creating many spinoff products, skipping steps, straining resources, becoming unfocused, and demoralizing team members. They do not consider the true long-term cost and impact of supporting each new product or market.

While discovering new markets, new ICPs, and new products to consider, make sure to ground the decisions in market research and validate your ideas before investing time and money into them.

Planning for an Exit

At this stage of the startup lifecycle, you are in a position to consider options for your future and the future of your company. While you may have considered your exit strategy in your Idea Stage business plan, as your company grows and the market changes, your exit strategy might also shift. It is almost impossible to predict how things will turn out and what opportunities will come your way. **Be open to different exit possibilities.**

At this point, you have multiple options to choose from, such as:

- Reinvest profits for organic growth.
- Cashflow the business and distribute profits to owners.
- Sell the business and cash out wholly or partially.
- Grow through acquisitions.
- Get acquired by or merge with another company.
- Go public.

If you decide to keep your startup, you might reinvest profits and continue to expand. Or you might distribute profits to owners. If you decide to start a new product, return to the initial stages of The 6 Startup Stages.

How to Value Your Company

If you decide to sell your company or acquire another company, it is important to understand the concept of business valuations.

When valuing a company, investors will look at several data points, such as:

- Competitive landscape and where your company fits
- Growth trends for gross revenue and net income
- Intellectual property assets such as patents, trademarks, and software
- Market growth potential
- Annual Recurring Revenue (ARR)
- Customer retention trends

Valuations are important to determining a company's future options. Many investors and founders like SaaS companies because subscription revenue is more predictable than other business models. Investors pay "higher multiples on SaaS ARR," making them easier to sell or get acquired. For example, a SaaS company with $1 million in ARR could, on average, see a $5 to $7 million purchase price because SaaS companies frequently sell at 5-7x of their ARR. Some SaaS companies have obtained up to 20x multiples on ARR but that is generally rare. There are always exceptions in very in-demand niches with companies that have amazing traction, growth, and founders. For a non-SaaS company, the ARR multiple is often 1–3x of annual revenue.

🎓 LESSONS FROM REAL SITUATIONS

Using highly unrealistic valuations could cause you to lose out on opportunities, partnerships, and funding.

We are often approached by startups who have unrealistic valuations while also having zero revenue, no product in the market, no past successes, and limited subject matter expertise.

As an example, we were approached by a Fintech startup founder who informed us their valuation was $20m and they asked us if we wanted to invest into their current round. No other investors had committed to invest at this valuation.

When we asked them about their valuation, they said this was a standard valuation for their industry at their stage. While the valuation was too high for us to engage, a much larger problem from our perspective was whether the founder had the ability to motivate others and set realistic expectations that would enable their company to succeed.

No matter what exit strategy you pursue, it is important to first understand your options, how each factor impacts your valuation, and how valuation decisions can impact other areas of launching and scaling your business.

What Not to Do During the Scale Stage

We have seen many companies try to grow too quickly. Before finding product-market fit, they will spend all their capital scaling up only to find they are going in the wrong direction. This causes them to waste a lot of time and money building and scaling a product without a solid customer base, leading to business failure.

TIP

"Don't get caught up wanting to announce you have raised millions of dollars in funding. It's better to learn from making less expensive mistakes, so you can preserve capital and equity."

Other things to avoid during the Scale Stage include:

- **Not focusing on the right KPIs.** Understanding what KPIs to focus on at each stage can get confusing. For example, founders may be focused on customer acquisition but lose sight of customer retention.

- **Underestimating the cost of scaling.** Maturing and scaling a business takes time, money, and expertise. Avoid scaling quickly with too many new hires or exhausting your current staff to the point of burnout.

- **Losing Focus.** Scaling can be overwhelming. Now is not the time to get distracted with too many different campaigns and strategies. Do not get too comfortable relying on processes that have brought your company to this stage, which may not apply to your current situation.

Scaling requires constant refinement of your business strategies, including analyzing data, staying curious about your ICP, and finding the right scaling strategies to help your business adapt and remain competitive.

Do Not Pass This Stage Until You

- ☐ Defined a strategy for maturing the business
- ☐ Defined a strategy to scale the business
- ☐ Extended the ICP
- ☐ Exited your business

How to use The 6 Startup Stages

I personally wish I had known about The 6 Startup Stages decades ago. I could have avoided many mistakes: skipping stages, jumping ahead, not validating ideas, or focusing on the wrong thing at the wrong time. I have learned a lot over the years from spending time talking to mentors, getting advice from coaches, reading business books, and attending business events.

Many people told us they have read a lot of business books but nothing brought everything together for them like The 6 Startup Stages.

No matter where you are in your startup journey, we know you can accelerate the success of your startup by leveraging **The 6 Startup Stages framework**. We have seen startups make almost every mistake at every stage of the process. We have also seen companies pivot, refine, adjust, and thrive. Customer needs are constantly shifting; therefore, you do not want to take too long moving between stages, or you might miss out on market opportunities.

Thank you for taking the time to read this book. We hope its guidance can help you turn your vision into a reality and help you build a successful, profitable company that can support others, do good in this world, and help you build a legacy for you and your family.

Bonus Tools and Content

As part of this book, additional free resources are provided to help you launch and scale your startup.

Get them here:

- Website Address:
  https://6startupstages.com/bonuscontent
- Content Password: *startupstagesbonus*

Get access to:

- Templates
- Worksheets
- Suggestions for online tools
- And more...

If you are not sure where you are in the startup journey, we recommend taking a free online quiz to get personalized results and recommendations. You can find the online quiz on the bonus content URL above.

Glossary

Launching and scaling a startup can be particularly challenging for non-technical founders who may not have a deep understanding of technology and related technology terms.

Below is a list of the top 100 glossary terms we believe every non-technical founder should get familiar with. These definitions include SaaS and non-SaaS startup terms. They cover many of the most important aspects, such as product development, customer acquisition, revenue generation, growth, retention, and valuation.

Whether you are planning to launch a startup, or you are already running one and want to improve your performance, this document will help you learn the vocabulary to prepare you for discussions with partners, employees, advisors, investors, and customers.

Definitions

1. **Churn Rate:** The percentage of customers who stop using a product or service within a given period of time, usually a month or a year. Churn rate is a measure of customer satisfaction and retention. Churn rate also affects the growth and profitability of a SaaS business. A high churn rate means that the business is losing customers faster than it is acquiring new ones. A low churn rate means that the business is retaining customers and increasing their loyalty.

2. **Agile Development:** A software development methodology that emphasizes collaboration, flexibility, and continuous improvement. Agile development involves breaking down large projects into smaller, manageable tasks, and delivering working software in short cycles called sprints. Agile development allows SaaS teams to adapt to changing customer needs, deliver value faster, and improve quality and efficiency.

3. **User Interface (UI):** The visual and interactive elements of a product, such as buttons, icons, menus, and colors. UI is a subset of UX and focuses on the aesthetics and functionality of the product. UI is important for SaaS products because it influences user engagement, satisfaction, and trust.

4. **Customer Acquisition:** The process of attracting and converting potential customers into paying customers. Customer acquisition is a key component of SaaS growth and requires a clear understanding of the target market, the value proposition, and the customer journey. Customer acquisition can be achieved through various channels, such as organic, paid, referral, or viral. Customer acquisition cost (CAC) is the average amount of money spent to acquire one customer.

5. **Customer Retention:** The process of keeping existing customers loyal and engaged with the product. Customer retention is a key component of SaaS growth and profitability, as it reduces churn, increases lifetime value, and generates

referrals. Customer retention can be achieved through various strategies, such as providing customer support, delivering value, soliciting feedback, and offering incentives. Customer retention rate (CRR) is the percentage of customers who remain customers over a given period of time.

6. **Customer Churn:** The process of losing existing customers to competitors or alternative solutions. Customer churn is a key challenge for SaaS businesses, as it reduces revenue, increases acquisition costs, and lowers customer satisfaction. Customer churn can be caused by various factors, such as poor product quality, lack of value, dissatisfaction, or switching costs. Customer churn rate (CCR) is the percentage of customers who stop being customers over a given period of time.

7. **Customer Lifetime Value (LTV):** The total amount of revenue generated by a customer over their entire relationship with the product. LTV is a key metric for SaaS businesses, as it measures a customer's long-term value and profitability. LTV can be calculated by multiplying the average revenue per user (ARPU) by the average customer lifespan. LTV can be increased by improving retention, increasing revenue, and reducing costs.

8. **Customer Acquisition Cost (CAC):** The average amount of money spent to acquire one customer. CAC is a key metric for SaaS businesses, as it measures the efficiency and effectiveness of the customer acquisition process. CAC can be calculated by dividing the total marketing and sales expenses by the number of customers acquired. CAC can be reduced by optimizing the acquisition channels, increasing conversion rates, and leveraging referrals.

9. **LTV/CAC Ratio:** The ratio of customer lifetime value (LTV) to customer acquisition cost (CAC). LTV/CAC ratio is a key metric for SaaS businesses, as it measures the return on investment (ROI) of the customer acquisition process. LTV/CAC ratio can be used to evaluate the scalability and sustainability of a SaaS business model. A higher LTV/CAC ratio indicates a higher ROI and a lower break-even point.

10. **Average Revenue Per User (ARPU):** The average amount of revenue generated by a user over a given period of time.

ARPU is a key metric for SaaS businesses, as it measures the monetization and growth potential of the user base. ARPU can be calculated by dividing the total revenue by the number of users. ARPU can be increased by upselling, cross-selling, or adding new features or services.

11. **Average Revenue Per Account (ARPA):** The average amount of revenue generated by an account over a given period of time. ARPA is a key metric for SaaS businesses, as it measures the monetization and growth potential of the account base. ARPA can be calculated by dividing the total revenue by the number of accounts. ARPA can be increased by expanding the account size, increasing the number of users per account, or adding new features or services.

12. **Annual Recurring Revenue (ARR):** The amount of revenue that a SaaS business expects to generate from its existing customers in one year. ARR is a key metric for SaaS businesses, as it measures the stability and predictability of the revenue stream. ARR can be calculated by multiplying the monthly recurring revenue (MRR) by 12. ARR can be increased by acquiring new customers, retaining existing customers, or increasing the revenue per customer.

13. **Monthly Recurring Revenue (MRR):** The amount of revenue that a SaaS business expects to generate from its existing customers in one month. MRR is a key metric for SaaS businesses, as it measures the stability and predictability of the revenue stream. MRR can be calculated by multiplying the average revenue per user (ARPU) by the number of users. MRR can be increased by acquiring new customers or increasing the revenue per customer.

14. **Annual Contract Value (ACV):** The amount of revenue that a SaaS business expects to generate from a single customer contract in one year. ACV is a key metric for SaaS businesses, as it measures the size and value of the customer contracts. ACV can be calculated by multiplying the monthly contract value by 12. ACV can be increased by upselling, cross-selling, or adding new features or services.

15. **Total Contract Value (TCV):** The amount of revenue that a SaaS business expects to generate from a single customer contract over its entire duration. TCV is a key metric for SaaS businesses, as it measures the size and value of the customer contracts. TCV can be calculated by multiplying the annual contract value (ACV) by the number of years in the contract. TCV can be increased by extending the contract length, upselling, cross-selling, or adding new features or services.

16. **Software-as-a-Service (SaaS):** A software delivery model that provides access to a cloud-based application over the internet. SaaS is a common business model for tech startups, as it offers several benefits, such as lower upfront costs, faster deployment, scalability, flexibility, and security. SaaS customers typically pay a subscription fee to use the software, which is hosted and maintained by the SaaS provider.

17. **Platform-as-a-Service (PaaS):** A software delivery model that provides access to a cloud-based platform that enables developers to create, deploy, and manage applications. PaaS is a common business model for tech startups, as it offers several benefits, such as lower development costs, faster innovation, scalability, flexibility, and security. PaaS customers typically pay a usage fee to use the platform, which is hosted and maintained by the PaaS provider.

18. **Infrastructure-as-a-Service (IaaS):** A software delivery model that provides access to a cloud-based infrastructure that enables users to store, process, and manage data. IaaS is a common business model for tech startups, as it offers several benefits, such as lower infrastructure costs, faster deployment, scalability, flexibility, and security. IaaS customers typically pay a usage fee to use the infrastructure, which is hosted and maintained by the IaaS provider.

19. **Pivot:** A significant adjustment in a startup's business model, product offering, or ICP, due to changes in the market, feedback from customers, or unexpected business challenges. The ability to pivot when needed, is critical to a startup's success. Before pivoting, a startup must carefully evaluate the effort and risks involved, to avoid unnecessary changes in focus.

20. **Augmented Reality (AR):** The technology that overlays digital information, such as images, sounds, or texts, onto the real-world environment, using devices such as smartphones, tablets, or glasses. AR is a common technology for tech startups, as it enables them to create products or services that can enhance the user's perception, interaction, or experience, such as gaming, education, or entertainment. AR customers typically pay a subscription or usage fee to use the AR, which is hosted and maintained by the AR provider.

21. **Virtual Reality (VR):** The technology that creates a simulated environment, such as a 3D world, that the user can interact with, using devices such as headsets, controllers, or gloves. VR is a common technology for tech startups, as it enables them to create products or services that can provide immersion, simulation, or escapism, such as gaming, education, or entertainment. VR customers typically pay a subscription or usage fee to use the VR, which is hosted and maintained by the VR provider.

22. **Pitch Deck:** A concise and visually compelling presentation created by startups to communicate their business concept to external investors to raise capital. A pitch deck should clearly and succinctly convey the startup's unique value proposition and essential aspects such as the market problem, the company's solution, traction, competitive landscape, team expertise, market opportunity, market potential, and use of funds.

23. **Growth Hacking:** The process of using data-driven, creative, and unconventional strategies and experiments to achieve rapid and sustainable growth for a product or service. Growth hacking is a common practice for tech startups, as it helps them acquire, retain, and monetize users, using minimal resources and costs. Growth hacking can involve various tactics, such as viral marketing, referral programs, gamification, or content marketing.

24. **Viral Marketing:** A marketing strategy that relies on users to spread the word about a product or service using social media, word-of-mouth, or other platforms. Viral marketing is

a common tactic for tech startups, as it helps them generate awareness, engagement, and adoption using minimal resources and costs. Viral marketing can involve various elements, such as emotional triggers, incentives, social proof, or network effects.

25. **Referral Program:** A marketing strategy that rewards users for inviting or referring other users to a product or service. Referral program is a common tactic for tech startups, as it helps them acquire, retain, and monetize users, using minimal resources and costs. Referral programs can involve various incentives, such as discounts, credits, free trials, or cash.

26. **Product Development:** The process of creating, designing, testing, and launching a new product or service that solves a problem or meets a need for a target market. Product development is a key activity for SaaS startups, as it helps them create value, differentiate themselves, and achieve product-market fit. Product development can involve various stages, such as ideation, prototyping, validation, iteration, or launch.

27. **Customer Acquisition:** The process of attracting and converting potential customers into paying customers. Customer acquisition is a key component of SaaS growth and requires a clear understanding of the target market, the value proposition, and the customer journey. Customer acquisition can be achieved through various channels, such as organic, paid, referral, or viral. Customer acquisition cost (CAC) is the average amount of money spent to acquire one customer.

28. **Revenue Generation:** The process of creating and capturing value from customers in exchange for a product or service. Revenue generation is a key component of SaaS profitability and sustainability, and requires a clear understanding of the revenue model, the pricing strategy, and the billing cycle. Revenue generation can be measured by various metrics, such as average revenue per user (ARPU), annual recurring revenue (ARR), or monthly recurring revenue (MRR).

29. **Growth:** The process of increasing the number, size, or value of customers, users, or accounts for a product or service. Growth

is a key objective and outcome for SaaS startups, as it helps them scale, compete, and create impact. Growth can be achieved by various strategies, such as growth hacking, viral marketing, or referral programs. Growth can be measured by various metrics, such as growth rate, customer retention rate, or customer lifetime value.

30. **Retention:** The process of keeping existing customers loyal and engaged with the product or service. Retention is a key component of SaaS growth and profitability, as it reduces churn, increases lifetime value, and generates referrals. Retention can be achieved through various strategies, such as providing customer support, delivering value, soliciting feedback, and offering incentives. Retention can be measured by various metrics, such as retention rate, churn rate, or net promoter score.

31. **Valuation:** The estimation of the worth or value of a SaaS startup based on various factors, such as revenue, growth, profitability, market size, or competitive advantage. Valuation is a key indicator and outcome for SaaS startups, as it helps them attract investors, raise funds, or exit. Valuations can be calculated by various methods, such as revenue multiples, discounted cash flow, or comparable transactions. Valuations can also be influenced by various factors, such as market conditions, investor sentiment, or negotiation skills.

32. **Minimum Viable Product (MVP):** The version of a product that has the minimum features and functionality required to test the product's value proposition and get feedback from early adopters. MVPs help SaaS founders validate their assumptions, learn from customers, and iterate quickly.

33. **User Experience (UX):** The overall impression and satisfaction that a user has when interacting with a product. UX encompasses all aspects of the product, such as design, usability, functionality, and performance. UX is important for SaaS products because it affects customer retention, loyalty, and word-of-mouth.

34. **Product-Market Fit:** The degree to which a product satisfies the needs and wants of a target market. Product-market fit

is a key factor for achieving product success and growth. A common way to measure product-market fit is by using the Net Promoter Score (NPS), which asks customers how likely they are to recommend the product to others. A high NPS indicates a strong product-market fit and a loyal customer base.

35. **Go-To-Market Strategy:** A Go-To-Market strategy, sometimes called GTM strategy, is how an organization plans to deliver their product to market and acquire customers. A GTM strategy defines the step-by-step plan for how to achieve this. The steps need to be tailored to each startup stage.

36. **Freemium:** A business model that offers a basic version of a product or service for free, and a premium version for a fee. Freemium is a common strategy for SaaS businesses, as it helps attract and acquire users, generate word-of-mouth, and create a loyal customer base. Freemium can be implemented in various ways, such as by limiting the features, functionality, users, or storage of the free version.

37. **Cloud Computing:** The delivery of computing services, such as software, platforms, infrastructure, storage, databases, and analytics, over the internet. Cloud computing is a common technology for tech startups, as it offers several benefits, such as lower capital expenses, faster innovation, scalability, flexibility, and security. Cloud computing customers typically pay a subscription or usage fee to use the cloud services, which are hosted and maintained by the cloud provider.

38. **Application Programming Interface (API):** A set of rules and protocols that defines how different software components or applications can communicate and exchange data. An API is a common technology for tech startups, as it enables them to integrate their products or services with other platforms, systems, or devices, and create new functionalities or features. API customers typically pay a subscription or usage fee to use the API, which is hosted and maintained by the API provider.

39. **Artificial Intelligence (AI):** The simulation of human intelligence processes, such as learning, reasoning, and decision-making, by machines or software. AI is a common technology for tech startups, as it enables them to create products or services that

can perform tasks that normally require human intelligence, such as speech recognition, image recognition, natural language processing, or machine learning. AI customers typically pay a subscription or usage fee to use the AI, which is hosted and maintained by the AI provider.

40. **Machine Learning (ML):** A subset of artificial intelligence that involves the use of algorithms and data to enable machines or software to learn from experience and improve their performance. ML is a common technology for tech startups, as it enables them to create products or services that can adapt to changing data, patterns, or behaviors, and provide insights or predictions. ML customers typically pay a subscription or usage fee to use the ML, which is hosted and maintained by the ML provider.

41. **Natural Language Processing (NLP):** A subset of artificial intelligence that involves the use of algorithms and data to enable machines or software to understand, analyze, and generate natural language, such as text or speech. NLP is a common technology for tech startups, as it enables them to create products or services that can interact with humans or other systems using natural language, such as chatbots, voice assistants, or sentiment analysis. NLP customers typically pay a subscription or usage fee to use the NLP, which is hosted and maintained by the NLP provider.

42. **Internet of Things (IoT):** The network of physical objects, such as devices, vehicles, or appliances, that are embedded with sensors, software, or connectivity, and can communicate and exchange data with other systems or platforms. IoT is a common technology for tech startups, as it enables them to create products or services that can provide connectivity, automation, and intelligence, such as smart home, smart city, or smart health. IoT customers typically pay a subscription or usage fee to use the IoT, which is hosted and maintained by the IoT provider.

43. **Net Promoter Score (NPS):** A measure of customer satisfaction and loyalty, based on the question: "How likely are you to recommend this product or service to a friend or colleague?"

NPS is calculated by subtracting the percentage of detractors (those who rate the product or service 6 or lower) from the percentage of promoters (those who rate the product or service 9 or 10). NPS can range from -100 to 100, with higher scores indicating higher satisfaction and loyalty.

44. **Customer Segmentation:** The process of dividing customers into groups based on their characteristics, behaviors, or needs. Customer segmentation helps SaaS businesses understand their customers better, tailor their products or services to their preferences, and optimize their marketing and sales strategies. Customer segmentation can be based on various criteria, such as demographics, psychographics, usage, or feedback.

45. **Software Development Life Cycle (SDLC):** The process of planning, designing, developing, testing, and deploying a software product or service. SDLC is a key activity for SaaS startups, as it helps them manage their product development, quality, and delivery. SDLC can involve various methodologies, such as waterfall, agile, or scrum.

46. **Customer Success:** The function of ensuring that customers achieve their desired outcomes and value from using a product or service. Customer success is a key component of SaaS growth and retention, as it helps customers adopt, use, and renew the product or service. Customer success can involve various activities, such as onboarding, training, support, feedback, or advocacy.

47. **Key Performance Indicator (KPI):** A measurable value that indicates how well a business is achieving its goals and objectives. KPIs are essential for SaaS businesses, as they help them monitor and improve their performance, growth, and profitability. KPIs can vary depending on the business model, stage, and strategy of the SaaS business. Some common KPIs for SaaS businesses are customer acquisition cost, customer lifetime value, monthly recurring revenue, churn rate, and net promoter score.

48. **Software Testing:** The process of verifying and validating that a software product or service meets the requirements and

expectations of the customers and stakeholders. Software testing is a vital activity for SaaS startups, as it helps them ensure the quality, functionality, and usability of their product or service. Software testing can involve various types, such as unit testing, integration testing, system testing, or user acceptance testing.

49. **Conversion Rate:** The percentage of visitors or users who take a desired action, such as signing up, subscribing, or buying. Conversion rate is a key metric for SaaS businesses, as it measures the effectiveness and efficiency of their marketing and sales efforts. Conversion rate can be calculated by dividing the number of conversions by the number of visitors or users. Conversion rate can be improved by various tactics, such as optimizing the landing page, offering a free trial, or providing social proof.

50. **Product Roadmap:** A strategic plan that outlines the vision, goals, and features of a product or service over time. Product roadmap is a key tool for SaaS startups, as it helps them communicate their product strategy, prioritize their product development, and align their stakeholders. Product roadmap can be presented in various formats, such as timeline, theme, or feature.

51. **Customer Feedback:** The opinions, suggestions, or complaints that customers share about a product or service. Customer feedback is a valuable source of information for SaaS businesses, as it helps them understand their customers' needs, expectations, and satisfaction. Customer feedback can be collected and analyzed by various methods, such as surveys, reviews, ratings, or interviews.

52. **Product Validation:** The process of testing and evaluating a product idea or hypothesis with real or potential customers before investing too much time or money into building it. Product validation helps SaaS founders avoid wasting resources on products that no one wants or needs, and instead focus on products that solve real problems and provide real value. Product validation can be done by various methods, such as interviews, surveys, landing pages, or prototypes.

53. **Scrum:** A framework that helps teams manage complex and collaborative projects, such as software development, by breaking them down into smaller and more manageable tasks, and delivering working products or services in short and fixed cycles called sprints. Scrum helps SaaS teams improve their productivity, quality, and efficiency, by providing a clear and structured way of planning, executing, and reviewing their work. Scrum can be implemented by following the roles, events, and artifacts defined by the Scrum Guide, such as the product owner, the scrum master, the development team, the sprint backlog, the sprint review, and the sprint retrospective.

54. **Pricing Strategy:** The method of setting the price of a product or service, based on various factors, such as the value proposition, the target market, the competitive landscape, and the business goals. Pricing strategy is a key component of SaaS revenue generation and profitability, as it affects the customer acquisition, retention, and lifetime value. Pricing strategy can be implemented by using various models, such as cost-based, value-based, or competition-based

55. **Market Size:** The total number of potential customers or users, and the total amount of revenue or demand, for a product or service in a specific market or industry. Market size is a key indicator and factor for SaaS growth and valuation, as it helps evaluate the opportunity, potential, and competition of a product or service.

56. **Competitive Advantage:** The unique value or benefit that a product or service provides to customers or users, that distinguishes it from other products or services in the same market or industry. Competitive advantage is a key factor for achieving product success and growth, as it helps attract, retain, and satisfy customers or users, and create a loyal customer base. Competitive advantage can be achieved by using various strategies, such as differentiation, cost leadership, or niche focus.

57. **Customer Persona:** A fictional representation of a typical or ideal customer or user, based on their characteristics, behaviors, needs, goals, and preferences. Customer personas

help SaaS businesses understand their customers or users better, tailor their products or services to their needs and wants, and optimize their marketing and sales strategies. Customer personas can be created and validated by using various methods, such as research, interviews, surveys, or analytics.

58. **Customer Journey:** The process of mapping and understanding the steps and stages that a customer or user goes through, from becoming aware of a product or service to using and renewing it, to becoming an advocate or a detractor. Customer journey helps SaaS businesses improve their customer experience, satisfaction, and loyalty, by identifying the pain points, needs, and expectations of their customers or users and providing solutions, value, and support along the way. Customer journey can be mapped and analyzed by using various tools, such as customer journey maps, touchpoints, or funnels.

59. **Customer Support:** The function of providing assistance and guidance to customers or users before, during, or after using a product or service. Customer support is a key component of SaaS retention and growth, as it helps customers or users solve their problems, answer their questions, and achieve their goals using the product or service. Customer support can be provided and measured by various channels, methods, and metrics, such as email, phone, chat, self-service, or social media, and response time, resolution time, or customer satisfaction score.

60. **Prototype:** A preliminary version of a product or service that demonstrates the core features and functionality without the full design or development. Prototypes help SaaS founders test their ideas, get feedback from potential customers, and refine their product or service.

61. **Market Research:** The process of gathering, analyzing, and interpreting information about a target market, such as the size, characteristics, needs, preferences, and behaviors of the potential customers. Market research helps SaaS founders validate their product or service ideas, identify

their market opportunities, and develop their marketing and sales strategies. Market research can be conducted using various methods, such as surveys, interviews, focus groups, or secondary data.

62. **Value Proposition:** The statement that summarizes the benefits and value that a product or service provides to the customers and how it differs from the competitors or alternatives. Value propositions help SaaS founders communicate their product or service's unique selling point, attract and retain customers, and increase conversions and revenue. Value propositions can be expressed using various formats, such as headlines, slogans, or pitches.

63. **Customer Advocacy:** The act of customers promoting or recommending a product or service to others, such as friends, family, or colleagues. Customer advocacy is a key component of SaaS growth and retention, as it helps generate word-of-mouth, referrals, and testimonials, and increase customer trust and loyalty. Customer advocacy can be encouraged by various strategies, such as providing incentives, creating communities, or sharing stories.

64. **Product Analytics:** The process of collecting, analyzing, and interpreting data about how customers use a product or service, such as the features, functionality, or performance. Product analytics helps SaaS founders measure and improve their product or service's value, usability, and satisfaction, and identify and prioritize areas for improvement or innovation.

65. **Design Thinking:** A creative and human-centered approach to solving problems and designing solutions that involves understanding the needs and preferences of the users, generating and testing ideas, and iterating and improving the outcomes. Design thinking is a common practice for SaaS founders, as it helps them create products or services that are desirable, feasible, and viable.

66. **Mockup:** A visual representation of how a product or service will look and function, without the actual coding or development. Mockups help SaaS founders communicate their product or service's features, functionality, and user interface,

and get feedback from potential customers, investors, or stakeholders.

67. **Wireframe:** A low-fidelity sketch or outline of the structure and layout of a product or service, without the details or design elements. Wireframes help SaaS founders plan and organize their product or service's content, navigation, and functionality, and define the scope and requirements of the project. Wireframes can be created using various tools, such as pencil and paper, or online software.

68. **Landing Page:** A web page that is designed to capture the attention and interest of visitors, and persuade them to take a specific action, such as signing up, subscribing, or buying. A landing page is a key component of SaaS marketing and sales, as it helps generate leads, conversions, and revenue. Landing pages can be optimized by using various elements, such as headlines, images, videos, testimonials, or calls to action.

69. **Lead:** A potential customer or user who has expressed interest in a product or service, by providing their contact information, such as name, email, or phone number. Leads are a key component of SaaS marketing and sales, as they help build a relationship, trust, and loyalty with the potential customer or user, and move them along the sales funnel. Leads can be generated by using various sources, such as landing pages, webinars, or social media.

70. **Sales Funnel:** The process of guiding and influencing potential customers or users from the initial stage of awareness, to the final stage of purchase, renewal, or advocacy. Sales funnels are a key component of SaaS marketing and sales, as it helps optimize the customer journey, increase conversions and revenue, and reduce churn and costs. Sales funnels can be divided into various stages, such as awareness, interest, consideration, decision, action, retention, or referral.

71. **Customer Relationship Management (CRM):** The system or software that helps manage and improve the interactions and relationships with existing and potential customers or users. CRMs are a key component of SaaS marketing and sales, as

it helps store and organize customer data, track and analyze customer behavior, automate and personalize communication, and increase customer satisfaction and loyalty. CRMs can be integrated with various tools, such as email, chat, or social media.

72. **Pre-Launch Marketing:** The process of building awareness, interest, and demand for a product or service before it is officially launched. Pre-launch marketing helps SaaS founders generate buzz, validate their product or service idea, and acquire early adopters. Pre-launch marketing can involve various tactics, such as creating a landing page, collecting email subscribers in a waitlist, offering a beta or a waitlist, or launching a referral program.

73. **Beta Testing:** The process of testing a product or service with a limited number of real or potential customers before it is released to the public. Beta testing helps SaaS founders identify and fix any bugs, errors, or issues, and get feedback and suggestions from the users. Beta testing can be done in various ways, such as closed beta, open beta, or public beta.

74. **Growth Rate:** The percentage change in the number, size, or value of customers, users, or accounts for a product or service over a given period of time, usually a month or a year. Growth rate is a key metric for SaaS businesses, as it measures the performance and potential of the product or service. Growth rate can be calculated by subtracting the previous period's value from the current period's value and dividing by the previous period's value. Growth rate can be increased by improving customer acquisition, retention, and revenue.

75. **Search Engine Optimization (SEO):** The process of improving the visibility and ranking of a website or web page on search engines for relevant keywords or phrases. SEO is a common strategy for SaaS businesses, as it helps them attract and acquire organic traffic, leads, and customers using minimal resources and costs. SEO can involve various techniques, such as keyword research, content creation, link building, or technical optimization.

76. **Search Engine Marketing (SEM):** The process of promoting and advertising a website or web page on search engines, such as Google or Bing, for relevant keywords or phrases. SEM is a common strategy for SaaS businesses, as it helps them attract and acquire paid traffic, leads, and customers using targeted and measurable campaigns. SEM can involve various methods, such as pay-per-click (PPC), cost-per-click (CPC), or cost-per-mille (CPM).

77. **Software Licensing:** The legal agreement that defines the rights and obligations of the software provider and the software user, such as the terms of use, the scope of access, the duration of service, or the payment model. Software licensing is a key component of SaaS revenue generation and sustainability, as it affects the customer acquisition, retention, and lifetime value. Software licensing can be implemented by using various models, such as perpetual, subscription, or usage-based.

78. **Software Integration:** The process of connecting and combining different software components or applications to create a unified and seamless system or platform. Software integration is a common technology for SaaS businesses, as it enables them to provide more value, functionality, and convenience to their customers or users, and create new products or services. Software integration can be achieved by using various tools, such as application programming interfaces (APIs), middleware, or connectors.

79. **Angel Investor:** An individual who provides financial support to early-stage startups, usually in exchange for equity or convertible debt. Angel investors are often wealthy, experienced, or influential people who can offer advice, connections, or mentorship to the founders. Angel investors typically invest smaller amounts than venture capitalists, ranging from a few thousand to a few million dollars.

80. **Seed Funding:** The initial stage of funding for a startup, usually used to develop the product, validate the market, and grow the user base. Seed funding can come from various sources, such as angel investors, venture capitalists, crowdfunding,

or bootstrapping. Seed funding typically ranges from a few thousand to a few million dollars, depending on the startup's valuation and traction.

81. **Series A Funding:** The second stage of funding for a startup, usually used to scale the product, expand the market, and grow the team. Series A funding usually comes from venture capitalists, who provide larger amounts of capital and expertise, in exchange for equity or preferred stock. Series A funding typically ranges from a few million to tens of millions of dollars, depending on the startup's revenue, growth, and potential.

82. **Series B Funding:** The third stage of funding for a startup, usually used to accelerate the growth, increase the market share, and improve the profitability of the product or service. Series B funding usually comes from venture capitalists, who provide even larger amounts of capital and guidance, in exchange for equity or preferred stock. Series B funding typically ranges from tens of millions to hundreds of millions of dollars, depending on the startup's performance, traction, and vision.

83. **Series C Funding:** The fourth stage of funding for a startup, usually used to expand the product or service to new markets or regions, acquire other businesses, or prepare for an exit. Series C funding usually comes from venture capitalists, private equity firms, or strategic partners, who provide very large amounts of capital and expertise, in exchange for equity or preferred stock. Series C funding typically ranges from hundreds of millions to billions of dollars, depending on the startup's valuation, revenue, and growth.

84. **Equity:** The ownership or stake that a founder, investor, or employee has in a startup, usually represented by shares or stock options. Equity is a key component of SaaS valuation and compensation, as it reflects the value and potential of the startup, and incentivizes the stakeholders to contribute to its success. Equity can be diluted or increased by various events, such as funding rounds, acquisitions, or employee stock option plans.

85. **Venture Capitalist (VC):** An individual or firm that provides financial support to startups, usually in exchange for equity or preferred stock. Venture capitalists are often experienced, influential, or wealthy people or organizations, who can offer advice, connections, or mentorship to the founders. Venture capitalists typically invest larger amounts than angel investors, ranging from a few million to billions of dollars, depending on the stage and potential of the startup.

86. **Bootstrapping:** The process of starting and growing a startup without external funding, relying on the founder's own savings, revenue, or personal loans. Bootstrapping is a common strategy for SaaS businesses, as it allows them to retain full control and ownership of their startup, and avoid dilution or dependency on investors. Bootstrapping can also be challenging, as it requires the founder to manage the cash flow, growth, and profitability of the startup, without much cushion or support.

87. **Exit:** The event or outcome that allows the founders, investors, or employees of a startup to realize the value of their equity, usually by selling their shares or stock options. An exit can occur through various ways, such as an acquisition, a merger, an initial public offering (IPO), or a secondary market. Exits are a key objective and indicator for SaaS startups, as it reflects the success and impact of the product or service, and provides a return on investment for the stakeholders.

88. **Initial Public Offering (IPO):** The process of offering the shares or stock of a startup to the public for the first time, usually on a stock exchange, such as Nasdaq or NYSE. IPO is a common way for SaaS startups to exit, as it allows them to raise capital, increase their visibility, and enhance their credibility. IPO can also be costly, complex, and risky, as it requires the startup to comply with various regulations, disclosures, and obligations.

89. **Family Office:** A private wealth management firm that serves one or more high-net-worth families, usually with assets of over $100 million. Family offices can provide various services, such as investment, estate, tax, or philanthropic planning. Family offices can also invest in startups, either directly or

through funds, and offer long-term, patient, and strategic capital.

90. **Private Equity (PE):** A type of investment that involves buying and selling shares or assets of private companies, usually with the aim of improving their performance, value, and profitability. Private equity investors are often large firms or funds that can provide capital, expertise, or strategic partnerships to the companies they invest in. Private equity investors typically invest larger amounts than venture capitalists, ranging from tens of millions to billions of dollars, depending on the size and maturity of the company.

91. **Cybersecurity:** The practice of protecting the data, systems, and networks of a product or service from unauthorized access, theft, or damage. Cybersecurity is a critical component of SaaS security and trust, as it helps prevent data breaches, cyberattacks, or malware infections and ensures the confidentiality, integrity, and availability of the product or service. Cybersecurity can involve various measures, such as encryption, authentication, firewalls, or antivirus software.

92. **Encryption:** The process of transforming the data of a product or service into a code or format that can only be read or accessed by authorized parties, using a secret key or algorithm. Encryption is a common technique for SaaS cybersecurity, as it helps protect the data from unauthorized access, theft, or modification, and ensure the privacy and security of the product or service. Encryption can be applied to various types of data, such as text, images, audio, or video.

93. **Authentication:** The process of verifying the identity and credentials of a user or device that accesses or uses a product or service, using a username, password, or other factors. Authentication is a common technique for SaaS cybersecurity, as it helps prevent unauthorized access, impersonation, or fraud, and ensure the accountability and legitimacy of the product or service. Authentication can involve various methods, such as single-factor, multi-factor, or biometric.

94. **Firewall:** A software or hardware device that monitors and controls the incoming and outgoing network traffic of a

product or service based on predefined rules or policies. A firewall is a common technique for SaaS cybersecurity, as it helps block or filter unwanted or malicious traffic, such as hackers, viruses, or worms, and ensures the performance and reliability of the product or service. Firewalls can be classified into various types, such as network, host, or cloud-based.

95. **Antivirus Software:** A software program that detects and removes viruses, malware, or other malicious software from a product or service using signatures, heuristics, or behavior analysis. Antivirus software is a common technique for SaaS cybersecurity, as it helps prevent or mitigate the damage or disruption caused by malicious software and ensures the functionality and safety of the product or service. Antivirus software can be installed on various devices, such as computers, smartphones, or servers.

96. **Data Breach:** The unauthorized or illegal access, disclosure, or theft of the data of a product or service by an internal or external party that compromises the security, privacy, or integrity of the data. A data breach is a major threat and challenge for SaaS cybersecurity, as it can result in financial losses, reputational damage, legal liabilities, or customer churn, and affect the trust and confidence of the product or service. Data breaches can be caused by various factors, such as human error, cyberattacks, or system failures.

97. **Cyberattack:** The deliberate or malicious attempt to harm, disrupt, or gain unauthorized access to the data, systems, or networks of a product or service, using various techniques, such as phishing, ransomware, denial-of-service, or brute force. Cyberattacks are a major threat and challenge for SaaS cybersecurity, as it can result in data breach, service outage, or extortion, and affect the availability and quality of the product or service. Cyberattacks can be motivated by various reasons, such as financial gain, espionage, or sabotage.

98. **Malware:** A type of malicious software that is designed to harm, disrupt, or gain unauthorized access to the data, systems, or networks of a product or service, using various techniques, such as viruses, worms, trojans, or spyware.

Malware is a common source and tool of cyberattacks and data breaches, as it can infect, damage, or steal the data, systems, or networks of the product or service, and affect the functionality and security of the product or service. Malware can be spread by various means, such as email, web, or removable media.

99. **Phishing:** A type of cyberattack that involves sending fraudulent or deceptive emails, messages, or websites, that impersonate a legitimate or trusted entity, such as a bank, a government, or a company, and trick the recipients into providing their personal or financial information, or clicking on a malicious link or attachment. Phishing is a common technique and threat for SaaS cybersecurity, as it can result in data breach, identity theft, or malware infection, and affect the privacy and trust of the product or service. Phishing can be targeted or generic, and can use various methods, such as spear phishing, whaling, or vishing.

100. **Ransomware:** A type of malware that encrypts the data, systems, or networks of a product or service, and demands a ransom from the owner or user, in exchange for the decryption key or the restoration of the access. Ransomware is a common technique and threat for SaaS cybersecurity, as it can result in data loss, service disruption, or extortion, and affect the availability and reliability of the product or service. Ransomware can be delivered by various means, such as phishing, drive-by downloads, or remote desktop protocol.

About the Authors

Jason Hishmeh

Partner | CTO | Tech Investor

- in https://linkedin.com/in/jasoncto
- X https://twitter.com/sjhmanhattan
- 🌐 https://6startupstages.com/jason

Jason is the co-author of "The 6 Startup Stages" book and The Startup Stages Guide.

He is a partner in Varyence.com and GetStartupFunding.com, and a limited partner/investor in the Luxembourg Gener8tor fund.

With over 20 years of experience in tech, his day-to-day responsibilities as a CTO and CISO include leading tech organizations for portfolio companies and commercial clients.

Jason enjoys speaking on topics related to Cybersecurity, Software Development, Cloud Infrastructure, Compliance, AI, and Startup Strategy.

He is an American citizen from New York City who spends much of his time in their EU and Lviv, Ukraine offices.

For media inquiries, please reach out to *media@6startupstages. com*.

About the Authors

Stas Chernychko

Partner | Head of Development | Tech Investor

🔗 https://linkedin.com/in/stascua/
🌐 https://6startupstages.com/stas

Stas is the co-author of "The 6 Startup Stages" book and The Startup Stages Guide.

He is a partner in Varyence.com and GetStartupFunding.com, and a limited partner/investor in the Luxembourg Gener8tor fund.

His day-to-day responsibilities as Head of Engineering include leading software engineering teams for portfolio companies and commercial clients.

He is a Ukrainian citizen residing in Lviv, Ukraine.

For media inquiries, please reach out to *media@6startupstages.com*.

About Varyence and Get Startup Funding

Jason and Stas lead a global company called **Varyence (Varyence.com)**. Varyence was founded in the United States, with offices and entities in the United States, the United Kingdom, Ukraine, and the European Union. Varyence helps startups, SMBs, and enterprise companies with seasoned technical leadership and product development teams. Additionally, Jason and Stas run a private fund called **Get Startup Funding (GetStartupFunding.com)** that invests in early-stage B2B SaaS startups.

With around 100 employees and decades of experience across startups and enterprise companies, we help founders successfully launch and scale their visions, create jobs for their communities, and build legacies for themselves and their families.

Our experience in design, application development, cybersecurity, cloud infrastructure, and compliance helps complement non-technical startup founders who are focused on sales, marketing, and customer acquisition.

As a company, we work with organizations in three ways:

- **Technical Outsourcing:** Varyence assists startups, SMBs and enterprise companies with design, development, innovation, scaling, operational efficiency, and cybersecurity.

- **Technical Cofounders and Business Partners:** Varyence invests in early-stage B2B SaaS startups, providing funding in the form of cash and product development services. As investors, we demonstrate our belief in our clients by helping back their projects with our own private capital, alongside other external investors.

- **Investment Accelerator:** Varyence is an LP (Limited Partner) in Gener8tor Luxembourg, an investment accelerator fund that invests up to €100,000 euros cash into multiple startups each year.

Raising Capital?

Every year, we connect with over 500 startups in various formats: helping review pitch presentations on investment panels, judging startups at pitch competitions, providing advice, referring them to helpful resources, and connecting them with potential investors.

We also directly invest every year into a limited number of startups from our private fund. We primarily focus on investments in early-stage B2B SaaS startups led by non-technical founders located in the US, Canada, the UK, and certain Western European countries.

If you have a great tech startup idea, are launching a tech startup, fundraising for a tech startup, or scaling an existing tech startup at any stage, we encourage you to reach out and stay in touch. If we cannot help you directly, we might be able to connect you with someone who can.

We also encourage you to check out and apply to our tech startup funding programs at https://getstartupfunding.com.

Keep in Touch

For regular tips and guidance, we encourage you to follow and connect with us on social media.

The 6 Startup Stages (6StartupStages.com)

THE **6 STARTUP**™ STAGES

Learn how non-technical founders can build profitable, scalable startups. Serial entrepreneurs share their insights, tips, and internal playbooks from startups they helped reach millions in annual recurring revenue or acquisition.

🌐 https://6startupstages.com/

in https://linkedin.com/company/6startupstages/

Varyence (Varyence.com)

varyence

Global software development company. In business for over 10 years, has around 100 employees, with offices and entities in the United States, the United Kingdom, Ukraine, and the European Union. Trusted by Startups, SMBs, and Enterprise companies.

🌐 https://varyence.com/

X https://twitter.com/varyence/

in https://linkedin.com/company/varyence/

Get Startup Funding (GetStartupFunding.com)

Global software development company. In business for over 10 years, has around 100 employees, with offices and entities in the United States, the United Kingdom, Ukraine, and the European Union. Reviews over 500 startups per year and invests into 3-5 per year.

🌐 https://getstartupfunding.com/

in https://linkedin.com/company/getstartupfunding/

Jason Hishmeh

Partner | CTO | Tech Investor

in https://linkedin.com/in/jasoncto

X https://twitter.com/sjhmanhattan

🌐 https://6startupstages.com/jason

Stas Chernychko

Partner | Head of Development | Tech Investor

in https://linkedin.com/in/stascua/

🌐 https://6startupstages.com/stas

Thank You

If you found this book helpful, we would appreciate it if you could share testimonies, feedback, and suggested topics for future writings. Please reach out to us via email at *authors@6startupstages.com*.

Additionally, we would appreciate it if you could add a great online review and share this book with others to help bring awareness to other startup founders who may also benefit from this information.

THE **6 STARTUP**™ STAGES

www.ingramcontent.com/pod-product-compliance
Lightning Source LLC
Chambersburg PA
CBHW071418210326
41597CB00020B/3563